Olivier Bruneau/Pere Grapí/Peter Heering/Sylvain Laubé/
Maria-Rosa Massa-Esteve/Thomas de Vittori (eds.)
Innovative Methods for Science Education:
History of Science, ICT and Inquiry Based Science Teaching

Olivier Bruneau / Pere Grapí / Peter Heering / Sylvain Laubé /
Maria-Rosa Massa-Esteve / Thomas de Vittori (eds.)

Innovative Methods for Science Education: History of Science, ICT and Inquiry Based Science Teaching

Frank & Timme

Verlag für wissenschaftliche Literatur

Cover illustration: Port militaire de Brest. Le pont tournant. © Archive de Brest

ISBN 978-3-86596-354-3

© Frank & Timme GmbH Verlag für wissenschaftliche Literatur
Berlin 2012. Alle Rechte vorbehalten.

Das Werk einschließlich aller Teile ist urheberrechtlich geschützt.
Jede Verwertung außerhalb der engen Grenzen des Urheberrechts-
gesetzes ist ohne Zustimmung des Verlags unzulässig und strafbar.
Das gilt insbesondere für Vervielfältigungen, Übersetzungen,
Mikroverfilmungen und die Einspeicherung und Verarbeitung in
elektronischen Systemen.

Herstellung durch das atelier eilenberger, Taucha bei Leipzig.
Printed in Germany.
Gedruckt auf säurefreiem, alterungsbeständigem Papier.

www.frank-timme.de

Contents

Preface	7

PART I

S. Laubé, O. Bruneau: **Inquiry Based Science Teaching and History of Science**	13
P. Heering: **Developing and evaluating visual materials on historical experiments for physics teachers: Considerations, Experiences, and Perspectives**	29
P. Grapí: **Using educational ICT to include history in science teaching and in science teacher training**	57
M. Massa-Esteve: **The Role of the History of Mathematics in Teacher Training Using ICT**	81
T. de Vittori: **Conceiving classroom IBST ICT EHST resources: practice analysis in mathematics**	109

PART II

S. Lawrence: **Inquiry based mathematics teaching and the history of mathematics in the English curriculum**	125
O. Bruneau, S. Laubé, T. de Vittori: **ICT and History of mathematics in the case of IBST**	145
H. Ferrière: **Remarks about ethical specificities of presenting online resources in history of biology for inquiry-based science teaching**	161
J. Gilliot, C. Pham-Nguyen, S. Garlatti, I. Rebai, S. Laubé: **Tackling Mobile & Pervasive Learning in IBST**	181
I. Kanellos: **Patrimonial Traditions Meet Educational Preoccupations: The Interpretive Shift of the Accessibility Requirement**	203

PART III

M. Bächtold, M. Guedj: **Towards a new strategy for teaching energy based on the history and philosophy of the concept of energy** 225

I. Guevara Casanova: **Historical contexts in mathematics curriculum for secondary school** 239

N. Cibulskaite: **Some peculiarities of mathematics teaching in Lithuanian basic school: computer, project, excursions, mathematics history** 251

S. Le Gars: **Introduction of a historical perspective in physics secondary school: replication and use of Galileo's experiments between theatre and modelisation** 275

C. Puig-Pla: **Suggestions for Introducing the History of Chinese Technology into Education** 287

R. Sucarrats Riera, A. Camós Cabecerán: **The application of the history of science and the use of the Internet as methodological tool in the subject of sciences of the contemporary world** 311

F. Romero Vallhonesta: **The Importance of Games of Chance at the Inception of Probability Theory.** 337

Author index 355

Preface

The project of this collective book results of several meetings since 2006 between European historians of science and technology. Regularly, both at national[1] and international[2] level, the six editors and most of the authors present in this publication organized symposia inside conferences about the role of history of science and technology in science education and teacher training.

In 2008-2010, the French group participated to the FP7 European group named "Mind The Gap" and the research time was ended in Brest, France, by a European Workshop entitled "Workshop "Mind the Gap" History of Science and Technology (HST): ICT Resources and Methods for Inquiry Based Science Teaching (IBST)"[3]. After the final dinner, we decided to publish a book in order to propose a "state of the art" that would point out the research activities that we leaded in France, Germany and Spain, in the domain.

We would like to underline in this preface two important facts that condition sustainably the research about HST and Education in future:

- The development of educational tools based on website, digital documents, collaborative work constitutes certainly a breaking point concerning teaching methods based on ICT as well the new interfaces between human being and machine (computer, mobile phone, enhanced reality, virtual world, etc.). A direct consequence is that new research problems in this area are strongly interdisciplinary and that mean collaborations with computer scientists.
- IBST is clearly a problem-based learning method which is considered at the European level as a good way in order to make more efficient the science education, to interest young people and to promote scientific culture in the society. History of science and technology con-

stitutes a very large "landscape" where to find examples to be used or adapted for IBST, but we would like to insist that it is important for the teacher, the teacher trainer (and of course the student) to keep safe the "gender" of HST when educational tools are designed.

Thus, the two objectives of the book are:

1. to enlighten and to discuss different research problems concerning HST and ICT, HST and IBST, HST and science education. In this way, it is dedicated to scholars.
2. to offer teacher and teacher trainer different ways to explore history of science and technology by using digital resources on-line, using new teaching method and to become more familiar with the method in HST.

The book is organized in three parts. The first one is a general approach proposed by the six editors. All the texts are based on the results of their research on HST and science education. In opening of this part, Laubé & Bruneau question the definition of a true inquiry based teaching and discuss the interest of "authentic" historical problems in science teaching. The development of new resources is one of the aspects explored by the following authors. In his text, Heering gives numerous examples on the early history of electromagnetism and raises important question about visual materials for physics teachers. In the same field, Grapí renders an account the building of an online course on history of science for in-service teachers and enlightens the potentialities of such a resource. In the field of mathematics, the contribution of Massa analyses the use of original sources in the classroom and the issues of teacher training in history of mathematics. The first part is ended by de Vittori's text in which the new didactical questions raised by the involvement of history in inquiry based classroom activities are examined.

The second part is dedicated to IBST and ICT design. In her text, Lawrence relates a work with original sources from the history of mathematics involving the video conferencing between different schools. The issues for teacher development and collaborative learning and teaching and the role the history of mathematics can play in this context is examined. This link between history of sciences and new technology is analyzed in the second text where Bruneau, Laubé & de Vittori render an account a large European project specifically on this topic. Then, Ferriére shows how the use of online resources in history of biology raises deep ethical questions, especially in a work on controversies. The last two texts of this second part of the book are from computer science specialists. Thus, with the light of this new field, Gilliot, Pham-Nguyen, Garlatti, Rebai & Laubé, explain how an inquiry based learning are fruitful for semantic web analyses. As for Kanellos, his contribution discusses the concept of accessibility to technical, scientific and generally cultural resources from a hermeneutical point of view. Let us know that henceforth, all the contributors of these last four papers are engaged in common research program on semantic web.

Varia papers constitute the third part. How can the history and philosophy of science be helpful in teaching the concept of energy? This question is the main entry point of Bächtold & Guedj's text in which they present the way they consider the connection between these two fields. All over Europe, the curricula are changing and mathematics education involves more and more history. Many ideas from the Catalonian situation are given by Guevara, and the case of Lithuania is very well described by Cibulskaite. History of science offers many resources and topics for those who want to introduce this perspective in science teaching. New examples are given in physics by Le Gars and in the rich text about Chinese mathematics by Puig-Pla. History of probabilities and its interest in teaching is the main topic of Romero Vallhonesta's contribution. Finally, as it is impossible to speak about new technology without mentioning Internet. Sucarrats & Camós explore how it can be considered as an interesting pedagogical tool for an historical approach in science teaching.

We would like to thank the international relations department of the Université d'Artois (France) for its financial support and all the persons or institutions that have made this project possible.

---

[1] for example, in Spain, a yearly colloquium is organized in November by the Catalan Society for the History of Science in Barcelona (see http://schct.iec.cat/) ; in France, several symposia were proposed by the ReForEHST group in the conferences of the French Society for the History os Science and Technology (see http://www.sfhst.org/) or in different University Institutes for Teacher Training (named IUFM) (see http://plates-formes.iufm.fr/ehst/rubrique.php3?id_rubrique=3)

[2] for example, see the website of the European Society for History of Science (see http://www.eshs.org/index.php?option=com_content&view=category&id=5&Itemid=81), the 11th Conference of the International History, Philosophy and Science Teaching Group (http://ihpst2011.eled.auth.gr/) or International Conference for the History of Science in Science Education (http://ichsse.ipcd.de/)

[3] see http://pahst.bretagne.iufm.fr/?p=84

PART I

Inquiry Based Science Teaching and History of Science

Sylvain Laubé[*] & Olivier Bruneau[+]

[*] Centre François Viète, Université de Bretagne Occidentale, Brest, France; sylvain.laube@univ-brest.fr

[+] Université de Lorraine, Laboratoire d'Histoire des Sciences et de Philosophie - Archives Henri Poincaré, UMR 7117, Nancy, F-54000, France; olivier.bruneau@univ-lorraine.fr

ABSTRACT: Inquiry Based Science Teaching can be considered as open "authentic" problem based learning in science. We will discuss in this paper the interest to study historical (and "authentic") problems in science in order to characterize a typology of problems to propose to the students. Some examples extracted from a paper review are proposed to illustrate our point of view.

Introduction

The *Centre Francois Viète (EA 1161)* in Brest (France) develops research works in the pluridisciplinar field of History of Science and Technology (HST), Heritage, Information and Communication based Technology (ICT), and also, in science education about the use of HST for Inquiry Based Science Teaching (IBST) and ICT tools for cultural mediation in science[1]. *Education* is taken here in the widest sense and concerns the place where knowledge in science is disseminated:

1) teaching and teacher training;
2) science and cultural mediation (museum, archives center, CCSTI[2], etc.).

In 2008-2010, our group (named *PaHST*) participated to the FP7 project "Mind the Gap" (n° 217725)[3] by producing a research report intitled "HST, ICT and IBST" (Laubé *et al*[4]) and organizing a European workshop « *History of Science and Technology: Resources and methods for Inquiry Based Science Teaching (IBST)* », in Brest, March 18th and 19th 2010.

This paper will summarize the principal elements of this report concerning IBST and HST. It is related to three others papers in this book that will develop the part concerning the role of ICT to develop HST resources for IBST: Bruneau *et al*, Gilliot *et al* and Kanellos.

Inquiry Based Science Teaching and History of Science

At the European level, the lack of students interest in science or in the scientific careers has led to a call for research projects in science education (the FP7 "Science in Society" program) and the publication of the Rocard Report about Science Education[5]. These recommendations promoted an evolution of teaching methods toward Inquiry Based Science Teaching (IBST) and requested some international comparisons.

In the FP7 "Mind The Gap" project, Inquiry Based Science Teaching was characterized by activities that pay attention to engaging students in:

- <u>authentic</u> and <u>problem based learning activities</u> where there may not be a correct answer
- a certain amount of experimental procedures, experiments and "hands on" activities, including searching for information
- self regulated learning sequences where student autonomy is emphasized
- discursive argumentation and communication with peers ("talking science")

Many of articles in Science Education Literature that we examined refer to the definition proposed by Linn *et al.*[6]: *"we define inquiry as engaging students in the intentional process of diagnosing problems, critiquing experiments, distinguishing alternatives, planning investigations, revising views, researching conjectures, searching for information, constructing models, debating with peers, communicating for diverse audiences, and forming coherent arguments"*.

Abd-el-Khalick *et al.*[7] summarized the results of a international symposium where two kind of inquiry appeared:

- *"<u>Inquiry as means</u> (or inquiry <u>in</u> science) refers to inquiry as an instructional approach intended to help students develop understandings of science content (i.e., content serves as an end or instructional outcome)"*

- "*Inquiry as ends* (or inquiry *about* science) refers to inquiry as an instructional outcome: Students learn to do inquiry in the context of science content and develop epistemological understandings about NOS and the development of scientific knowledge, as well as relevant inquiry skills (e.g., identifying problems, generating research questions, designing and conducting investigations, and formulating, communicating, and defending hypotheses, models, and explanations)."[8]

Several descriptors were chosen to characterize the role of inquiry in science education: "These include scientific processes; scientific method; experimental approach; problem solving; conceiving problems, formulating hypotheses, designing experiments, gathering and analyzing data, and drawing conclusions; deriving conceptual understandings; examining the limitations of scientific explanations; methodological strategies; knowledge as "temporary truths;" practical work; finding and exploring questions; independent thinking; creative inventing abilities; and hands-on activities"[9]. Furthermore, the authors specified that: "This set of descriptors also focuses our attention on the need to distinguish within our curricula what it is we wish to be the goals of science education (e.g., content, process, NOS) and how an inquiry approach to science education can (or cannot) help achieve these goals"[10].

In the FP7 "Mind The Gap" Project, we pointed out that an historical approach of "authentic" problems in science is helpful to characterize what is Inquiry in Science and what kind of problems have to be solved. From the studies about scientific theories, concepts creation, and the way how experiments are elaborated and analyzed, historians of science showed that science does not only consist in final results: the processes take also an important part of knowledge elaboration, like scholars hesitations between two, or more, models, how scientists create experiments, collect data, discuss the results, etc. In each field of science (mathematics, physics, biology and Earth science), history of science gives interesting and authentic examples that show the complexity and the richness of knowledge construction. There is no doubt that these historical data are useful to describe and analyze the investigation process. Inquiry Based Science Teaching has to be aware of this, as it underlines its own main concepts. The FP7 research report was based on examples in mathematics, biology and physics. We will here illustrate the methods by focusing about the Galileo's works that constitute a set of suitable examples in order to show what inquiry is in authentic and historical scientific situations.

Galileo's works as an example of authentic scientific inquiry

Galileo showed a remarkably appreciation for the proper relationship between mathematics, theoretical physics, and experimental physics. *Sidereus Nuncius*[11] is thus an authentic and historical example that allows *inquiry in science* to be understood. *Sidereus Nuncius* is a short treatise published in Latin in March 1610. It was the first scientific treatise based on observations made through a telescope. It contains the results of Galileo's early observations of the Moon, the stars, and the moons of Jupiter. It gives arguments against the Aristotelian "Weltanschauung" and in favour of the Copernican view where the Sun is in centre of the world. Galileo received in 1609 a report concerning a telescope constructed by a Dutchman and decide "*to inquire into the principle of the telescope*". He succeeded in constructing an instrument so good that the objects appeared magnified thirty times nearer. He explained the method to construct the telescope, some elements about the theory and physical principles and the way to use it. Concerning the moon, Galileo observed that the darker part makes it appear covered with spots. He draws several sketches in order to describe the observations. In the last portion of Sidereus Nuncius, Galileo reported the observation (made between January 7th and March 2nd 1610) of the motion of four stars that appeared to form a straight line of stars near Jupiter with illustrations of the relative positions of Jupiter and the stars.

The discovery of spots on the moon surface and of the fours stars moving near Jupiter constitute two problems that was not solved inside the Aristotelian theory. Hypothesis and models were stated in order to explain what he observed. About the Moon, Galileo explains that the darker regions are low-lying areas and brighter regions are covered with mountains. From this hypothesis, he calculated that the lunar mountains were at least four Italian miles in height: *"We are therefore left to conclude that it is clear that the prominences of the Moon are loftier than those of the Earth"*. About the four stars near Jupiter, he shows that the movement doesn't belong to Jupiter (as it was first believed), but to the stars (named then Medicean Planets). *"It can be a matter of doubt to no one that they perform their revolutions about this planet, while at the same time they all accomplish together orbits of twelve years' length about the centre of the world. [...] the revolutions of the satellites which describe the smallest circles round Jupiter are the most rapid"*. The Jupiter system with his four Medicean Planets appear here as a Kepler's model.

Galileo is Copernican and all his discoveries are used as "arguments" against Aristotelians and *"remove the scruples of those who can tolerate the revolution of the planets round the Sun in the Copernican system, yet are so disturbed by the motion of one Moon about the Earth, while both accomplish an orbit of a year's length about the Sun, that they consider that this theory of the constitution of the universe must be upset as impossible ; for now we have not one planet only revolving about another, while both traverse a vast orbit about the Sun, but our sense of sight presents to us four satellites circling about Jupiter, like the Moon about the Earth, while the whole system travels over a mighty orbit about the Sun in the space of twelve years."*

The Sidereal Messenger, is totally included in the context of the controversy between Aristotelians and Copernicans. What is inquiry in the Sidereal Messenger? We can see three types of problems. First, the technological questions are linked to scientific instruments. How to construct an instrument? What theory is used to explain it? How to use it? Second, the enigma has to be well posed and to be solved by explanatory models. But modelling requires collecting data (observations, measurements, etc.) in order to obtain reference data as input for the construction/discussion of the model. Those reference data are reports, sketches, tables, numerical data, etc. Third, these enigmas are included in a larger theoretical controversy and the constructed models are participating here to a debate that will lead to the "Copernican Revolution".

This example shows that history of science could good furnish material or references for an "authentic" inquiry-based learning in science and technology. Each point can be regarded as *problem solving situations* and for this side of the scientific activity the history cannot be ignored.

As result of the FP7 research, we proposed then to consider IBST as Open Problem Based Science Teaching (in a set of activities where student autonomy is emphasized) about: 1) collecting data 2) stating hypothesis, 3) testing hypothesis, 4) experimentation/hands on, 5) modelling, 6) results evaluation, 7) argumentative communication, 8) scientific language.

A review to illustrate HST/IBST activities as Open Problems

From a paper review on the topic "HST and IBST", we selected some papers in order to give concrete examples.

Collecting Data

Dolphin[12] explains that collecting data is an important moment in IBST – first for the teacher who has to review good resources - when we are looking for a theoretical explanation and a dynamic model (here the tectonic model of earth). So, the author collects – with his pupils - data in the past (historical representation of the earth: historical texts, textbooks, patterns, maps…), data in his classroom (or maybe in museum and environment for fossils, rocks, photos…) and first visual representations of his pupils. This example is a global approach of investigation in science that integrates historical and epistemological approach in the same time.

P. Clément[13] talks about collecting different kind of cells in different times (and method of collecting from and coloring cells of plants and animals), about some instruments (optic and electron microscopes), about different ways to present (in museum, in university and school), to show, describe and symbolize (photos, texts, draws, patterns) and their consequences for understanding other phenomenon (like cellular differentiation or epithelial level of organization in an animal organism).

P. Mihas[14] establishes an account of some experiments which come from history to develop ideas on refraction. One of them starts with the editing of data with the Ptolemy's method and he notices that *"Ptolemy's Refraction experiment results can be compared with students' results. This can be done by asking students to plot their results with Ptolemy's results in Excel and try to find a relation between the angles. (...) This exercise helped students appreciate the value of planning for an experiment.* »[15]

Thus, a great diversity of data can be mobilized via historical IBST, and there are not only texts but also maps, photos, fossils, etc.

Stating hypothesis

From Ptolemy's and their proper data collection, Mihas' pupils compare their and propose a refraction law (the Ptolemy's law): "These results were presented to the students to compare with their own results. The students recognized easily the implied relation. The students also questioned the results. The author asked the students to tabulate their own results and to compare with Ptolemy's."[16]

Testing hypothesis

Students test hypothesis in order to validate, to reject or to amend this one. Testing induces a critical reflection and allows that some different hypotheses are possible. Former hypothesis depend on the finiteness of knowledge of students like historical scientist: "mathematicians" inadequate knowledge about the convergence of infinite series in the 17th and 18th century was also brought up in the class. One of the assignments was to sum up the divergent infinite series "1 + 1 - 1 + 1 - 1 + 1 ..." *wherein students were given three contradictory but seemingly reasonable answers in history and asked to select the correct one. This episode reflected the unsound foundation of calculus at that time and the potential fallibility of superficial intuition.*"[17]. Rudge and Howe notice also the importance to test reflection: *"At this time, the instructor encourages students to discuss their answers to questions about nature of science issues that were given to them on the slip of paper during their group work. (...) The first probe invites students to indirectly consider what is commonly referred to as the subjective (theory-laden) nature of science. A conception that it often held by students is associated with a naïve-inductivist perspective, which holds that students believe scientists inevitably all come to similar conclusions when examining the same data."*[18] The example developed by Dolphin is also very attractive because he states and tests (with his pupils) different "essential questions" (*"Questions that are not answerable with finality in a brief sentence but are used to stimulate tought, to provoke inquiry, and to spark more questions"*[19]) then some hypothesis. They are about the structure, composition, temperature, history and dynamism of Earth. But, in the same time, he does not explain why several models of earth have been proposed by scientists in different time: why several hypothesis have been well considered and some other have been immediately forgotten for different reasons (like pragmatics or empirics reasons: observations by miners by example, but we could also talk about the belief of the existence of Hell in the middle of Earth to explain the models with fire or "lavas ocean" under our feet...).

Experimentation/ hands on

In the IBST activity, the central point is experimentation. In our literature review, this part is present and relatively well explained. The Dolphin's work is also a good example in our case: his students have to build some model scale to understand the different point of view in history of Geology (to know the age, the structure and the dynamism model of Earth). In this case, there is a real hands-on experimentation with lot of sort of things that are chosen by professor: sponges, balloon, paper (so the autonomy of students – point 8 – is not so bigger than we could believe at the beginning)... In others cases, the selected investigation is more a "literature search" than the rest. But, it is also a particularity of HOS to permit to select a kind of investigation especially when replications of experimentation are impossible (for many reasons as ethic reason or the expensive cost of a replication...) or in the case of mathematics, the main source is a written one. For instance, Liu explains how pupils experiment many ways of demonstration: *"instead of introducing limited concepts at the outset, students were asked to prove or explain the area of a circle is [proportional to] r^2 by using basic mathematics at the middle-school level, followed by the introduction of historical approaches used by Archimedes, Japanese mathematician Seki Kowa, and ancient Chinese mathematician Liu Hui. This problem-solving activity aimed to increase students' experiences and understanding of infinitely partitioning processes and the sum of infinite vanishing quantities."*[20].

One can find in history of science a lot of kind of experiments and some of them can be reproduced relatively easily in classrooms. For instance, Riess et al. reproduce the Galileo's inclined plane experiments: *"One of the best known experiments with respect to the discussion of free fall is the inclined plane experiment that was published by Galileo Galilei in 1638. This experiment has been analysed by the Oldenburg group with the replication method. Currently, we are working on teaching material that will give access to our experiences with this set-up for teachers and students. Moreover, we are reconstructing a demonstration apparatus developed in the 18th century to teach free fall and the superposition principle. Historically contextualised, this experiments."*[21]

Modelling

Dolphin is convinced that modelling and discussing about modelling is a central part and allow pupils to develop critical views about science: *"An important part of the contextualized approach is the use, discussion, and critique of models. Models play an important role in teaching science content and teaching about the nature of science. (...) One major challenge was taking concepts which represent some of the major discoveries or paradigm shifts that occurred during the evolution of the theory of plate tectonics and developing different modes of representation for them. Because students often confuse a simplified model for its target, they need to be exposed to many different modes of representation in order to facilitate enrichment of their mental models and their understanding of the concept. (...) By organizing the models historically and allowing students to discuss and debate them, the process of how science really works is itself modeled. (...) My motivation is for students to sharpen their own critical thinking skills by separating themselves from their own mental models and analyzing those models for strengths and limitations. Critical assessment of models by students is encouraged with the use of model analysis worksheets used throughout the entire course of study.*"[22]

Results evaluation

Assessment of results is probably more effective when students have understood how they were built in the past. The historical and contextualized approach allows finding the nature and complexity of the evidence in the demonstration. Outcome evaluation is not only about the soundness of arguments and evidence but also about their efficiency and construction. Glenn Dolphin, N. Gericke and M. Hagberg[23] work show that the results evaluation is more effective when students can compare them with historical results.

Moreover, one can consider the evaluation as a good challenge and source for pupils of pleasure. For instance, Koponen and Mäntylä state about using 19th-century physical experiments that: "*Such experiments can, nevertheless, be used to help students' conceptualisation in support of learning. Students can still have the satisfaction of participating in creating the knowledge for themselves although it now becomes strongly guided by the teacher and constrained by empirical observations.*"[24]

Argumentative communication

Liu pays attention to the fact problem-solving activities product social interaction and participate to reinforce argumentative communication: "*In addition to the history-based curriculum, students were situated in a dynamic problem-solving setting. As aforementioned, several historical problems were used in serving the purpose of bringing out students' curiosity and the desire to think. Students demonstrating more elaborated thinking were invited to share their ideas and approaches. All questions and comments from peers were welcomed for developing critical thinking. Following whole classroom discussion, the class then learned about the relevant historical background and mathematicians' approaches. For more challenging problems, such as Napier's original logarithm and Leibniz's tractrix problem, students worked in groups to increase social interaction and motivate higher-order thinking.*"[25] Rudge and Howe put out that an inquiry-based teaching must include discussions and debates. And at this stage, "*In the remaining time for this class, the instructor answers students' general questions about the conclusions they reached for the Uganda data or answers questions students have about other curious aspects of the mystery patient already examined in a prior class. The instructor reminds students to write a reflective diary entry about their experiences in the class with the expectation that they turn in their diary entries at the beginning of the next class. Students are also reminded to consider the diary probe when writing about their experiences in the class.*"[26]

Scientific language

Introduce non-werstern history of science shows how to do science in the other way especially when pupils are not sensible to European culture. For instance, Wang declares: "*At least, the history of ancient Chinese mathematics can show us another way to do mathematics, which is very different from the western tradition and the system in modern textbooks. This way can enable us to think, to learn and to teach mathematics in a more interesting way, one that is easier to understand and more related to the reality than is now the case.*"[27]

As conclusion of this paragraph, we would like to highlight once again a citation of Abd-el-Khalick F. et al[28]: "Thus, instead of thinking of a generalized image of inquiry in science education and assuming it will allow achieving multiple goals, <u>it might be more useful to think of several images that are intimately linked with small clusters of valuable instructional outcomes.</u> What is needed is a sort of a multidimensional heuristic that defines a space of outcomes, which would facilitate discourse and streamline communication about images of inquiry between players within any educational setting (e.g., policymakers, curriculum theorists and developers, administrators, teachers, teacher educators, and students), such that the likelihood of impacting actual classroom practices related to inquiry is substantially increased. [...]One dimension could include the types of knowledge and understandings that Duschl refers to, that is, conceptual, problem solving, social, and epistemic. Another dimension could include a range of inquiry-related activities, such as, problem-posing; designing investigations; collecting or accessing data; generating, testing, and refining models and explanations; communicating and negotiating assertions; reflecting; and extending questions and solutions. A third dimension could include a range of (the necessarily reductionistic but nonetheless crucial) skills, such as mathematical, linguistic, manipulative, and cognitive and metacognitive skills, needed to meaningfully engage in inquiry at one level or another. A fourth dimension could comprise a range of spheres, including personal, social, cultural, and ethical, with which any of the aforementioned outcomes could interface. When navigating through this four-dimensional space, one could think of the elements on each dimension either as possible outcomes of, or as prerequisites for meaningful engagement in, inquiry-based science education. The former would help conceive and place more emphasis on inquiry as means (inquiry as teaching approach), while the latter thinking would help gauge the level at which students could engage in inquiry and help emphasize inquiry as ends (inquiry as an instructional outcome".

Conclusions

The first conclusion that we would propose is that the paper review showed that the research field about IBST is very active. The questions linked to the use of history of science in science education are studied at the European level by historians as it is shown for example by the symposium "HST & Education" organized inside the two last Conference of the European Society of History of Science in Vienna (2008) and Barcelona (2010)[29].

As historians of science and technology, we do consider IBST as a Open Problem Based Science Teaching. We have shown that HST could contrib-

ute to produce authentic pedagogical problems well adapted to IBST. We think now that we need a European network in order to:
- publish on-line resources in history of science in different european language and well-adapted to education
- produce research results about "HST & Education"

The second conclusion concerns the research field about resources in HST with ICT tools for IBST and cultural mediation. The research field concerned three scholar communities: Computer Science, Science Education and History of Science. As it is said in the European network of excellence Kaleidoscope[30], research questions about technology-enhanced learning systems concerns the ability to reuse learning resources (learning objects, tools and services) from large repositories, to take into account the context and to allow dynamic adaptation to different learners, contexts and uses based on substantial advances in pedagogical theories and knowledge models (Balacheff[31]). The design and engineering of learning systems about IBST or cultural mediation with resources in HST must be considered as a "big" interdisciplinary research problem requiring the integration of different scientific approaches from computer science, pedagogical and/or didactical theories, education, history of science, etc. The design process leads to an artifact - the learning system - based on different scientific approaches which are related to different theories – for instance, activity theory, theory of didactic situations, computer-based theories, etc. Consequently, it is crucial to establish the relationships between theories, models and artifacts to ensure the traceability and the interpretation of phenomena related to the use of artifacts (Tchounikine *et al*[32]).

Thus, we think that a major point for the future is to work on adaptive technology-enhanced learning systems for cultural mediation using a problem-based learning approach and represented by IBST scenarios[33]. The goal of scenarios is to describe the learning and tutoring activities to acquire some knowledge domain and know-how to solve a particular problem. A scenario may depend on several dimensions which describe different learning situations: the learning domain (course topic), the learner (his know-how and knowledge levels), the tutor/teacher, the learning and tutoring activities (their typology, organization and coordination), the resources (documents, communication tools, technical tools, etc.), the activity distribution among learners, teachers and computers, the learning "procedures" according to a particular school/institution/ university and the didactical/pedagogical environment.

In other words, dimensions are closely related: changing one dimension may lead to the change of others. For instance the learning activities have to change according to the learner know-how and knowledge levels for a given knowledge domain. In other words, their typology, organization and coordination change to deal with these dimensions. *Adaptive technology-enhanced learning systems* compute on the fly the delivered courses from distributed data resources, according to the current context and the learner's needs. The resource reusability has to rely on resource interoperability at syntactic and semantic level. At semantic level, resources are described by semantic metadata and their corresponding ontologies[34]. These ontologies can be used to formalize at knowledge level the different required models of learning systems: user models (student, teacher, visitor of a museum) models, domain model (i.e. the gender of digital documents in HST[35], IBST), context model, scenario models, pedagogical and/or didactical models, adaptation models and rules, etc. New software architectures are necessary to use learning system models based on ontologies and to support dynamic adaptation and context awareness.

The results and the interest of The European "Mind the Gap" workshop organized in Brest[36] in March 2010 was to show three axis in order to develop HST technology-enhanced learning systems for IBST in the future:

- The necessity to develop Web 3.0 ICT tools[37] in order to share the resources at the european level
- The necessity to publish historical digital documents for science education at the european level and, thus, to propose a translation in the

different european languages of the fondamental historical texts or documents in science
- The interest for historians of science and computer science researchers to work together about ICT and innovation[38]

---

[1] For example, see the website « Ressources en histoire des sciences et techniques pour la formation des maîtres »: http://plates-formes.iufm.fr/ressources-ehst/spip.php?rubrique18.
The item « Histoire des techniques » was realized with the help of Brest Archives and educative service of the National museum of Marine.
[2] Centre de Culture Scientifique Technique et Industrielle (Cultural Center dedicated to Science, Technology and Industry)
[3] http://uv-net.uio.no/mind-the-gap/index.html
[4] Laubé S., Bruneau O., Ferrière H., de Vittori T. "History of Science, ICT and IBST", Deliverable 5.4, FP7 Project "Mind the Gap" n°217725
[5] http://ec.europa.eu/research/science-society/index.cfm?fuseaction=public.topic&id=1100
[6] Linn, M. C., Clark, D., Slotta, J. D. (2003). WISE design for knowledge integration. *Science education*, 87, 517-538.
[7] Abd-el-Khalick F, Boujaoude S, Duschl R, et al. "Inquiry in Science Education: International Perspectives". *Science Education*. 2004; 88(3), 397-419.
[8] Abd-el-Khalick *et al.*, *op. cit*, p. 398.
[9] *Id.*, p.411-2.
[10] *Id*, p. 412.
[11] Galileo, Galilei, *Sidereus Nuncius*, http://www.intratext.com/IXT/LAT0892/_P3.HTM (Original version in Latin); Galileo, Galilei, *The Sidereal Messenger*, http://www.archive.org/details/siderealmessenge80gali (English translation, 1880); Galilei, Galileo, *Le messager céleste: contenant toutes les nouvelles découvertes qui ont ete faites dans les astres depuis l'invention de la lunette d'approche*, A Paris: Claude Blageart ..., et Laurent d'Houry ..., 1681, http://fermi.imss.fi.it/rd/bd?lng=en&collezioni=galileiana
[12] Dolphin G. Evolution of the Theory of the Earth: A Contextualized Approach for Teaching the History of the Theory of Plate Tectonics to Ninth Grade Students. *Science & Education*. 2009; 18(3-4):425-441.
[13] Clément P. Introducing the Cell Concept with both Animal and Plant Cells: A Historical and Didactic Approach. *Science & Education*. 2006; 16(3-5):423-440.
[14] Mihas P. Developing Ideas of Refraction, Lenses and Rainbow Through the Use of Historical Resources. *Science & Education*. 2006; 17(7):751-777.
[15] Mihas, *op. cit*, p. 755.
[16] Mihas, *op. cit.*, p 758.
[17] Liu P. History as a platform for developing college students' epistemological beliefs of mathematics, *International Journal of Science and Mathematics Education*. 2008;7(3):473-499, p. 477-8.
[18] Rudge and Howe, *op. cit.*, p. 571.
[19] Dolphin, *op. cit.*, p.428.

[20] Liu, *op. cit.*, p. 477.
[21] Riess F, Heering P, Nawrath D, Education P. Reconstructing Galileo's Inclined Plane Experiments for Teaching Purposes. In: *Eighth International History, Philosophy, Sociology & Science Teaching Conference*. Leeds; 2005:1-10, p. 1.
[22] Dolphin, *op. cit.*, p. 427-8.
[23] Gericke NM, Hagberg M. Definition of historical models of gene function and their relation to students' understanding of genetics. *Science & Education*. 2006; 16(7-8):849-881.
[24] Koponen IT, Mantyla T. Generative Role of Experiments in Physics and in Teaching Physics: A Suggestion for Epistemological Reconstruction ". *Science & Education*. 2006; 15(1):31-54, p.51.
[25] Liu, *op. cit.*, p.477-8.
[26] Rudge and Howe, *op. cit.*, p. 572.
[27] Wang, *op. cit.*, p. 639.
[28] Abd-el-Khalick F., Boujaoude S, Duschl R, et al. Inquiry in Science Education: International Perspectives. *Science Education*. 2004; 88(3), *op. cit.*, p. 415
[29] Fourth ESHS Conference, Barcelona,
http://conf.ifit.uni-
klu.ac.at/eshs/images/M_images/PDFs/suss.%20liste.%206.%20sept.pdf
[30] http://www.intermedia.uio.no/display/Im2/Kaleidoscope
[31] Balacheff, N. (2006). "10 issues to think about the future of research on TEL." Les Cahiers Leibniz, Kaleidoscope Research Report (147).
http://www-didactique.imag.fr/Balacheff/TextesDivers/Future%20of%20TEL.pdf
[32] Tchounikine, P. and Al. (2004). Platon-1: quelques dimensions pour l'analyse des travaux de recherche en conception d'EIAH. *Rapport de l'action spécifique "Fondements théoriques et méthodologiques de la conception des EIAH"*, Département STIC, CNRS.
http://telearn.noe-kaleidoscope.org/warehouse/Tchounikine_2004.pdf
[33] Laubé S., Garlatti S., Tetcheung J.-L. (2008) "A scenario model based on anthropology of didactics for Inquiry-Based Science Teaching". *International Journal of Advanced Media and Communication*, april 2008, vol. 2, n° 2, pp. 191-208.
[34] about ontologies for computer science, see: http://semanticweb.org/wiki/Main_Page
[35] Laubé S. "Modélisation des documents numériques pour l'histoire des techniques: une perspective de recherche", *Documents pour l'histoire des techniques*, n° 18, décembre 2009, pp. 37-41. http://assoc.secdhte.fr/wp-content/uploads/2010/01/dht18_laube.pdf
[36] http://pahst.bretagne.iufm.fr/?p=84
[37] http://ec.europa.eu/information_society/eeurope/i2010/invest_innov/index_en.htm
[38] http://ec.europa.eu/information_society/tl/research/documents/ict-rdi-strategy.pdf

Developing and evaluating visual materials on historical experiments for physics teachers: Considerations, Experiences, and Perspectives

Peter Heering

Institute for Physics and Chemistry and its Didactics, Universität Flensburg; D- 24943 Flensburg; Germany; peter.heering@uni-flensburg.de

ABSTRACT: This paper discusses an attempt to develop visual materials for physics teachers on experiments from the early history of electromagnetism. In several respects, this attempt has to be interpreted as a pilot study: the number of materials that were developed is few, the field and period are very narrow, and the evaluation of the developed materials can serve only as a pilot study as the number of teachers evaluated has been very limited. Yet, it appears to be relevant to discuss some of the ideas that were conceptually important for the development of the materials. In this respect, also the results of the pilot study of the evaluation appear to be meaningful as they seem to strengthen some of our considerations.

Introduction

Various accounts have attempted to implement history of science in general and the history of scientific experimentation in particular in science education (see in particular the publications of the Oldenburg group led by Falk Rieß, but also Achilles (1996), Cavicchi (2003), Kipnis (1993), and Teichmann (1979)). These approaches aim at enabling students to redo historical experiments, either with reconstructions that are created according to the available source information, or with reconstructions that are based on the working principle of the instruments. Evidently, the experiences made in these approaches show that such a approach is beneficial in science education. However, a problem appears to be the availability of the instruments that are necessary for such an approach. The approach I am going to discuss in this paper goes in a completely different direction. I will discuss the development and evaluation of visual materials that are meant to train and support teachers who are going to teach the physics of electrodynamics through its history. The materials were developed in the STeT-Project (Science Teacher e-Training).[1] The project started in summer 2006 and was completed in late 2008.[2] It aimed at training and supporting science teachers by developing materials that were intended of an interactive website. Through the training as well as the as the accessibility of the materials on the website, the teachers were meant to be motivated and enabled to teach science through its history. The content of the materials was limited to electrodynamics from the 19th century. As the materials were intended for being used by the teachers, they were not necessarily to be suited for a direct use by the students. The materials were meant to address either primary or lower secondary school teachers. Consequently, the content and depth of discussion within the materials varies significantly.

Among the materials that were developed are texts such as biographies. These materials are intended to serve as background information to the teacher. Consequently these texts are fairly dense, and they contain substantial details so that teachers can choose where to focus in their teaching. Other materials are intended to enable role plays in the classroom; still others are computer simulations and videos of experiments. The two latter materials are going to be central to this paper. Before discussing them in greater detail, it appears necessary to make a few remarks on the meaning of these materials in the entire conception of the project.

Experiments were meant to play an important role in the conception for two reasons: First of all, they are highly relevant for the development of the field of electrodynamics. Even though the development of theories and concepts plays an important role (see e.g. Blondel (1982), Jungnickel & McCormmach 1998), recent historiography has demonstrated that also experimentation is relevant and to a certain extent theory independent, and cannot be neglected (see e.g. Meya (1990), Steinle (2005)). Yet, in some sense this is only a minor part of the reason: As the materials are intended for becoming a basis for teachers' with respect to their practice in the science classroom, one has to take into consideration also educational aspects. In science education, experiments are still dominant, even though some publications seem to indicate that experiments are not the panacea for the science classroom. It appears that experiments need to be well chosen and embedded in other activities in order to be beneficial for science education. These aspects had to be taken into consideration. Consequently, the materials on experiments developed in the project were related to stories and other materials on the website.

In order to offer teachers a variety of methodological opportunities, the website had to fulfil different purposes with respect to the use of experiments in the classroom: On the one hand it should provide ideas for experiments that can be either demonstrated by the teacher or that can be carried out by the students. On the other hand, we also developed videos that show reconstructed experiments and which are meant to be shown in teaching situations. To a certain extent, it remained open how the materials were to be used in the science classroom. However, when developing these materials we held certain ideas about how to support science education with experiments that are presented through videos.

Some general considerations about ICT and historical experiments

One might of course consider that every historical experiment that can be re-enacted could also be videotaped and shown in the classroom. Probably this is possible, yet, one should question whether this is useful from a didactical point of view. A key problem appears to be that the own activity of the students shows the most efficient growth of competence. However, watching a video is a situation where students are kept passive, in other words they consume what is presented without any own involvement. Consequently, one has to decide which experiment is shown as a video in a classroom and – perhaps even more important – which one is not to be shown but serves as a basis for an experimental activity of the students.

Yet, there is also another aspect that should not be underestimated: Historical experiments were performed in different manners and according to different standards.[3] As the aim of the entire approach of education through the history of a field does not just focus on an efficient accumulation of content knowledge by the students but aims also at aspects such as the development of an understanding of the nature of science in general and science as a cultural activity in particular, it is crucial that the experiments remain historically contextualized. In this respect, an approach that aims to create a somewhat symmetrical situation between the historical and the educational appears to be necessary. Consequently, in cases where experiments were shown as demonstrations in the historical situation, a video might be the appropriate manner to represent this situation in a classroom. However, in cases where direct interaction is crucial (e.g. measurements or experiments with physiological effects), a video might not be adequate as the skills of the experimenter – which are necessary for performing the experiment –cannot be represented in such a video. Consequently, either other materials or ideas of how to actually perform these experiments should be used.

Yet, there is another factor which needs to be taken into consideration. Even though a teacher might intend to carry out historical experiments in her or his classroom, he or she needs the required equipment. However, this equipment is not necessarily part of the standard collection a school has available.[4] In this respect, the argumentation made in the paragraph above appears to be somewhat academic in the sense that it does not take into consideration the (material, but also other) conditions under which teachers are to work in their classroom. Consequently, it might be necessary for one teacher to use a video in a classroom whilst a teacher at another school has different working conditions and can thus use the same video as a starting point for developing an activity her or his students are about to make.

Another central topic with respect to the videos was the language: As the videos were intended to be used in e-training programmes of different countries, there appeared to be at least two options. One could either prepare versions of the videos for each partner country (and potentially also an English version). Or one could prepare videos with a general language – potentially English – and eventually use subtitles. In the end, we decided to go for a third option: we decided to prepare the videos without any text and just have the necessary information for the teachers on the website. In this manner, we could enable teachers of all participating countries to use similar materials but at the same time adapt them to their needs.

Examples from the history of electromagnetism

Before discussing the visual materials that were developed in the project, it might be useful to sketch very briefly the central figures and experiments we took as a basis for developing these materials. The discovery and subsequent discussion of galvanic electricity can be seen as the beginning of the history of electromagnetism. Consequently, the experiments by Galvani on frogs as well as his conceptual development of animal electricity serve as potential starting point. At the same time, Volta's work on contact electricity – as a rival conception to the one of Galvani – and his development of the electrical battery could be seen in the same context. A second 'couple' can be formed by Oersted and his work on magnetic effect of an electrical current and Ampère's subsequent work on electromagnetism. And finally, we decided to include Faraday and his development of the rotating apparatus, a device that turned electromagnetic interaction into a permanent mechanical motion. This work can be seen in the context of the development of the electrical motor. Jacobi's motor is the second part of this storyline that goes from the development stabilization, and communication of a physical effect to its first technical application.[5] Instead of examining the historical meaning of these experiments any further, the ideas with respect to the materials that were developed shall be discussed.

Realisation

Galvani's experiments of frogs whose legs start to move when brought in contact with two different materials are certainly a good starting point as well as in some sense potentially fascinating to the students. Yet, there are of course ethical reasons that make it impossible to carry out these experiments in a classroom or in a science lab for educational purposes. In this respect, I strongly oppose a respective suggestion by Kipnis: "If possible, do these experiments in a laboratory setting." (Kipnis 2001, 268). Even though the strong visual effect might be considered as useful for the introduction into the field, it is certainly not possible (and at least in Germany also illegal) to kill a frog in order to do electrical experiments with its legs for educational purposes. This example might serve as a good example where videos can be used in a classroom, and one can certainly question whether killing one single frog in order to prepare this video is that problematic. However, at least from my perspective, I consider this approach also to be inappropriate. Moreover, such a video might produce a negative attitude of the students towards physics as they can see it as an example of how physicists treat animals.

Fig. 1: Screenshot from the video on Galvani's experiments.[6]

Instead of preparing a video of this experiment, we decided to use computer simulations (Fig. 1). They offer insights that are comparable to a video demonstration without being ethically that problematic. These videos are meant to be shown in the classroom, additionally, plates from Galvani's publication as well as a summary of his experiments were provided. Using the simulations and the plates were also meant to enable teacher to discuss why such experiments are no longer done by scientists. In this respect, changing standards and thus cultural influences on scientific practices can be discussed on a completely different level than usual.

Like in the case of Galvani, we also decided to prepare no video of Volta's experiments on the pile. In some of his experiment, he also used frogs' legs as indicators of the electrical action; here, we also used simulations for visualization. In other experiments, Volta used the own body in order to feel the electrical action. This cannot be shown in video, except if one asks a person making this experience to give a description. Yet, as the materials were to be used in different countries, we had the intention of preparing videos that are without spoken or written information that would require a translation. Instead, the additional material that gave the background information of the experiment was also used as giving additional explanations about the experimental procedures.

However, the general limitations of such a description pose another problem: it is difficult to verbalize the experiences of closing a voltaic pile with the own body, and it is even more difficult to understand such a description. Therefore we decided that besides the videos of the simulation, we would also prepare a visual manual of how to build a 'crown of cups' (Fig. 2), a device that is similar to the pile and was also described in Volta's first publication (Volta 1800).[7]

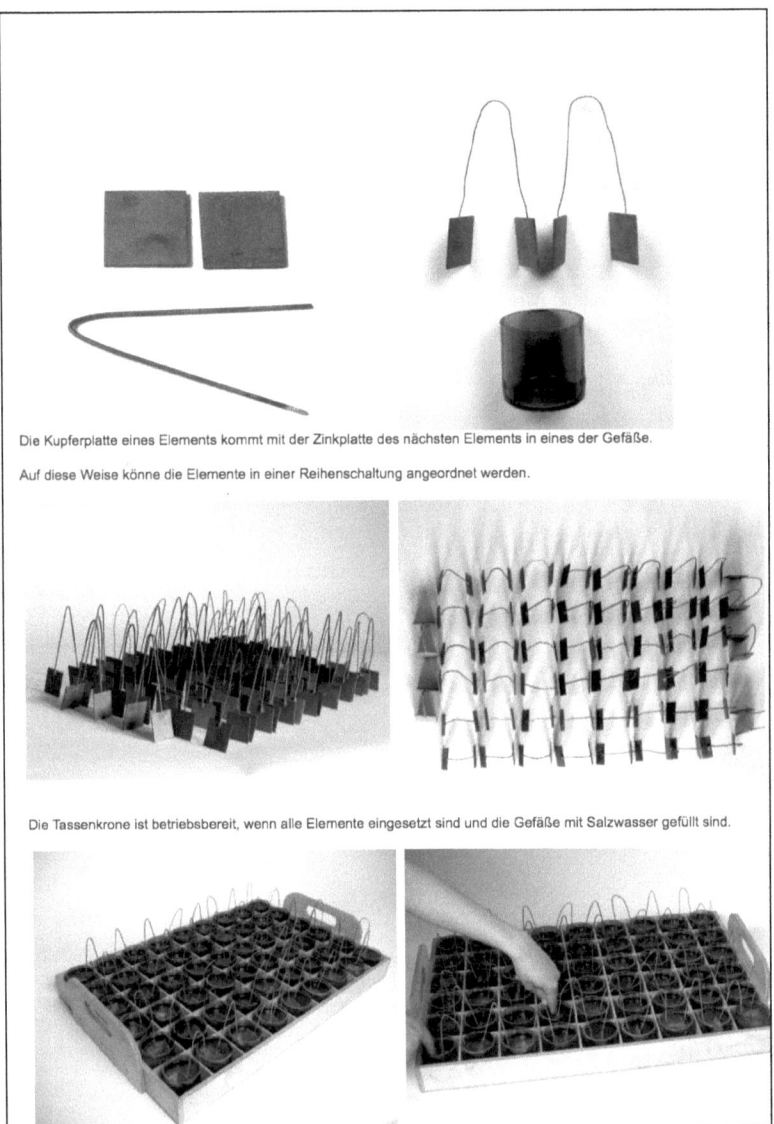

Fig. 2: **Screenshots with the visual manual for building the ‚crown of cups'**[8]

Whilst this first case focuses on the very beginning of research in the field of galvanic electricity, the second one deals with the beginning of research in electromagnetism. Here, Oersted's publication (1820) can be taken as a clear starting point.[9] Remarkably enough, Oersted published his findings in Latin, a procedure that was fairly uncommon in 19th century physics. A crucial aspect of the experiment turned out to be the initial orientation of the needle with respect to the direction of the current, and the orientation of the needle's deflection. As this was (at least for most contemporary researchers) the unexpected part of Oersted's findings, he probably aimed at avoiding any confusion and potential criticism due to resulting difficulties in replicating his findings. For that purpose, it was important to have a proper understanding of the orientation of the needle and its deflection with respect to the current in the wire. In order to avoid difficulties and errors in translating these details, Oersted used a language that was still familiar to all scientists. This printed account was not sent to journals but to several researchers whom Oersted knew, some of them published a translated account in the respective journals of their country (see Steinle 2005).

With respect to establishing his findings, Oersted was certainly successful; his experiment was replicated immediately all over Europe. However, he was still criticized by his contemporaries for having found the effect accidentally – a perspective that can be understood as a criticism towards the Romantic Movement of which he had been associated with (Brain 2007). This criticism still serves as a basis for suggestions how the effect can be taught with an historical perspective (Kipnis 2005). However, there is a convincing alternative as it can be shown that Oersted actually had a conceptual approach that enabled him to predict the effects he then observed (Martins 2003).

The key aspect in the experiment is the question of where to place the needle and how it is displaced when the electrical circuit is closed. From a technical point of view, an educational version of the experiment is simple to be realized – one needs only a strong battery (like a car battery), a magnetic needle and some wire. However, creating regularity in the displacement of the needle is more complicated (until one has understood the regularity, then it is of course simple, too). In this respect, it appears that the procedural aspect of the experiment is the crucial part. Consequently we decided not to prepare a video but to give more background information about the historical context of the experiment instead.

Things are different with the work of Ampère. He started his experiments immediately after the emergence of Oersted's findings in Paris. In the following months, Ampère carried out series of experiments and read several papers to the Paris' Academy of Sciences, in the end, he was able to show and describe the electromagnetic interaction of two currents (Blondel 1982, Steinle 2005).

With respect to the development of the materials, particularly Steinle's description was extremely helpful. He distinguishes between two types of experimentation, exploratory and theory-oriented. Simplified, it can be said that exploratory experimentation is found when a researcher aims to develop a conceptual understanding. The respective experiments are characterized by a theoretical openness, however, parameters are varied in a systematic manner. In theory-oriented experimentation, the researcher has a conceptual understanding that is used to create the experiment. In case of electromagnetic interaction, Ampère started with analyzing the interaction between a magnet and a current – this is still very much related to Oersted's experiment. He used experimental designs where he placed a bar magnet that was suspended with a thread at different orientations towards a spiral or a coil which was part of the circuit that was closed (Fig. 3). The spiral or the coil respectively was also suspended; Ampère was thus able to show the mutual attraction and repulsion between the magnet and the current. In his next experiments, he replaced the magnet with a second coil or spiral. In doing so, he was able to observe and analyse the electromagnetic interaction between two currents (Fig. 4). At the end of this period of his experimentation he designed his current balance (Fig 5). Thus, this device can be seen as a materialisation of Ampère's conceptual development. It is no longer that open to manipulation or modification but a stable device that shows exactly two effects (the attraction or repulsion between two electrical currents, depending on their orientation). As this instrument is designed for a demonstration, it appeared also well suited for the demonstration in the classroom in a video. The lack of options to manipulate the instruments together with the aspect that is not simple to build such a device makes it from our perspective highly unlikely that teachers will aim at reconstructing this device.[10]

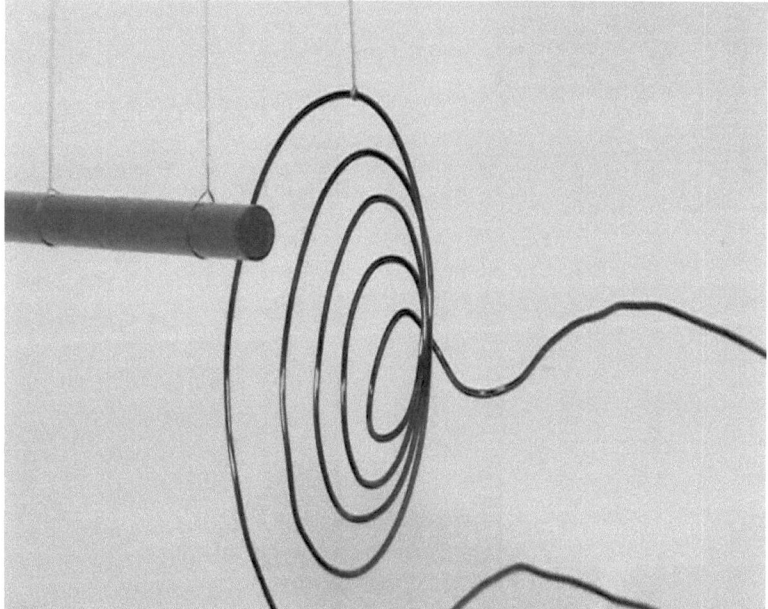

Fig. 3: Suspended Magnet and Spiral before closing the circuit

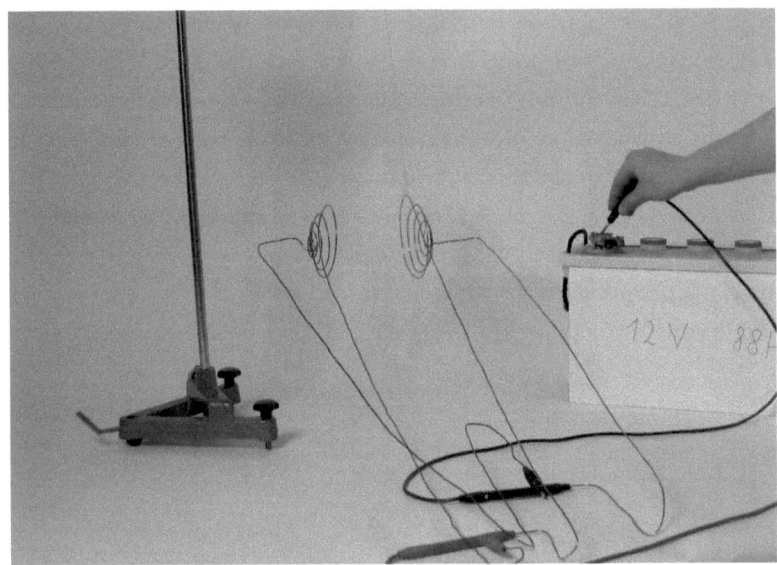

Fig. 4: Repulsion between two spirals

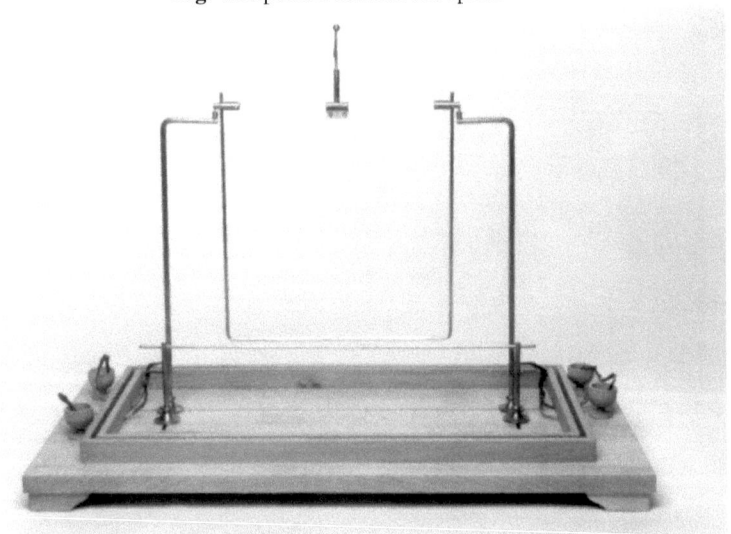

Fig. 5: Reconstruction of Ampère's current balance. Foto W. Golletz

Whilst the current balance was designed to demonstrate acquired knowledge, this is different with respect to the experiment with magnets, coils, and spirals. Here, the experiments were more open to the development of a theory. Therefore, these experiments seem to have a larger potential for teaching situations.[11] As a result, our idea would be to have the experiments carried out in the classroom – either as a demonstration by the teacher or in form of a student activity. In order to facilitate both options, videos were made that show the variety of arrangements and the resulting effects. However, it is evident that these experiments can be carried out easily in the classroom instead of showing the videos.[12]

The final example we developed in this project focused on the technical application of the electromagnetism, namely the electrical motor. Faraday had been working on electromagnetic interaction and developed a device which could be used to transform electromagnetic interaction into a permanent magnetic movement. This device, which he called rotation apparatus, consists of a glass cylinder which has a cork at the top and at the bottom. The cylinder is filled to about half its height with mercury. A soft iron pole is pushed through the cork at the bottom and terminates over the surface of the mercury, a thin metal wire that ends in an eye through the cork at the top. Into the eye, another wire is hooked which ends in the mercury. At the beginning of the experiment, a magnet is held next to the soft iron pole. The pole and the wire outside of the top cork are connected with the poles of a battery. Consequently, the electrical circuit goes from the battery to the soft iron pole, the mercury, the two wires, and back to the battery. As a result there is a current going through the wire which is inserted into the mercury and which is in the magnetic field that is produced by touching the soft iron with a magnet. As a result, a force acts onto the wire which starts to rotate around the magnet. Faraday built several versions of this rotating apparatus, among them one where the wire was fixed and the magnet moveable, consequently, the magnet rotated around the wire.[13] Like in the case of Ampère's current balance, the rotating apparatus served as a material demonstration of the knowledge Faraday had achieved, consequently, we envisioned that this instrument can be presented in the classroom with the help of a movie.

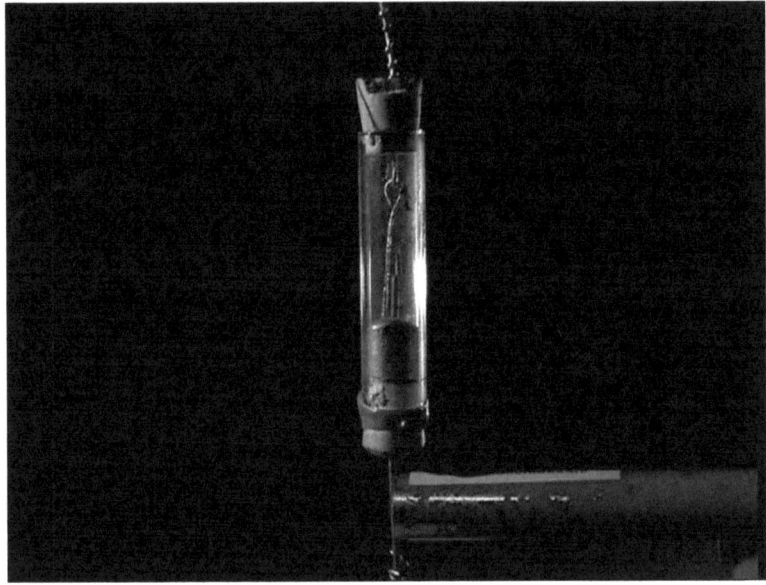

Fig. 6: Reconstruction of Faraday's rotating apparatus

Even though not even Faraday considered this device as an electrical motor, it is sometimes mentioned in the literature as the first example. However, if one takes as a criterion that a motor is not just demonstrating the physical working principle but actually can be applied as a technological object, then the first electrical motor was built by the St. Petersburg professor of physics Moritz Hermann von Jacobi. It consists of eight pairs of U-shaped electromagnets, four of which form the stator, the other four the rotator (see Fig. 7, the wooden disc carries the magnets which form the rotor, the rectangular frames the ones which form the stator). Jacobi used his motor to propel a boat on the river Newa. A crucial detail of the motor is the commutator (Fig. 8), a device that changes the polarity of the electric current for one set of magnets, thus enabling a permanent mechanical rotation. It consists of two pairs of brass discs, each pair is connected to one end of each of the electromagnets that form the rotator. In all four discs are (non-conducting) wooden segments inserted. These insertions replace one eighth of the disc each, and are placed with a difference of 45° in the two discs that form a pair. Consequently, one disc has a contact to the brass lever which is lying on it and which is connected to the coil that forms the electromagnet, the other does not. When the north poles of the stator are directly opposite of the south poles of the rotator, the polarities of the magnets on the rotator change, due to the inertia, the rotator moves on and the two alike magnets repel each other whilst the unlike magnets are attracting each other. Consequently, as there are four pairs of magnets, the polarity changes eight times in a full circle.[14]

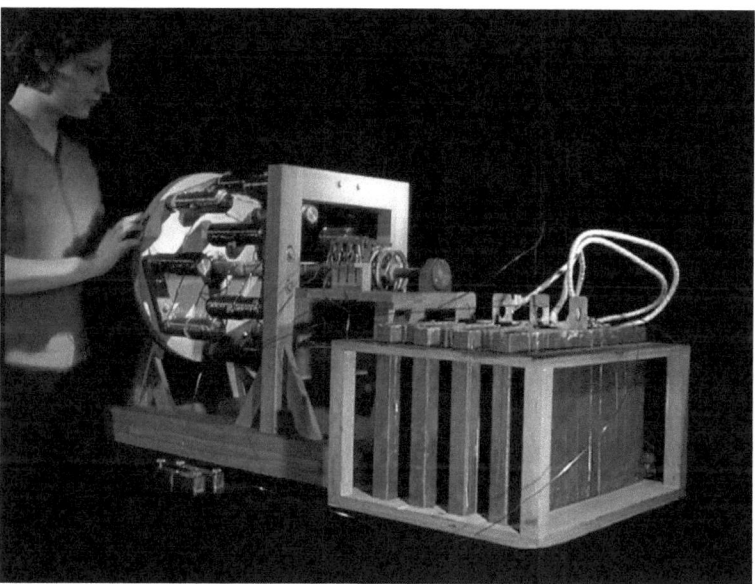

Fig. 7: Reconstruction of Jacobi's electrical motor

Fig. 8: Reconstruction of the commutator of Jacobi's electrical motor

Like in the case of the rotating apparatus, we decided to prepare a video showing the operation of the motor. First of all, there is not much interaction that is possible for students with the device (even though measurements to determine the efficiency could be realized and help to understand why the electrical motor was established as a technological device only some decades later). Secondly, the motor is heavy and expensive, thus it is extremely unlikely that students will be able to build their own version of this device.[15]

Evaluation

Having developed the materials, it was of course an issue to evaluate whether the ideas on which these materials were based are reasonable and whether these materials appear to be useful for school science teaching. As the materials were developed for teachers, and they are the ones who decide whether they are going to use these materials, we considered them to be the relevant target group for the evaluation. The evaluation was realized with three groups of teachers who were trained on the job. Most of them had finished university within the last two years, all of them were teaching physics in a Gymnasium in Lower Saxony. The teachers were asked to examine the website and its content. Additionally, some of the teachers participated in a one-day workshop on historical experiments in science education. The examples in the workshop were taken from the history of electrostatics. Consequently, the materials the teachers evaluated in the following were unfamiliar to them prior to the visit of the website.

All teachers were given a written questionnaire with open as well as multiple choice questions. All in all we send the questionnaire to 42 teachers; 31 of them responded. In the open question, some suggestions were made of how to improve the materials. To give but one example: One teacher suggested having the original Latin text of Oersted's paper included in the website so that the physics teacher can collaborate with the Latin teacher – this idea has meanwhile been realized.

Summarizing the general response to the materials (Diagram 1), one could say that the half of the teachers indicated that they considered the materials as useful for their teaching. This is of course also related to the fact that the topics (electromagnetic interaction, electrical motors in general and Jacobi's in particular) were compulsory in the teaching.

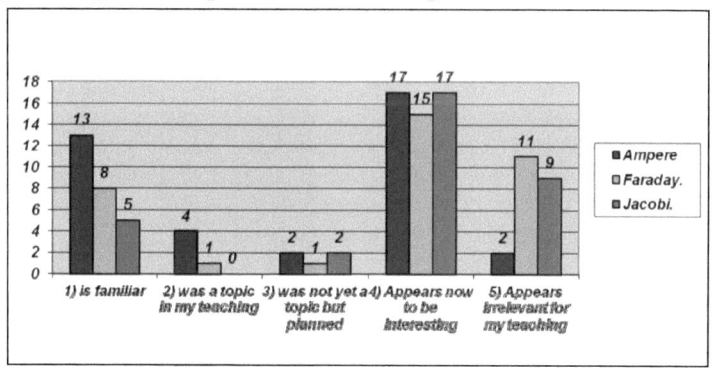

Diagram 1: Evaluation of the materials' relevance to teaching

Labelling the materials as useful for their teaching can also be seen as an indication that the materials were adequate with respect to their conceptual understanding of their own teaching. However, even though this response is not to be seen as an indication that they are going to teach these topics through their history, at least the materials appear relevant to them on a very basic level.

It may seem astonishing that one third of the teachers indicated that the materials on Faraday and the Jacobi motor as being irrelevant for their teaching. However, as these topics were not familiar to most teachers and they are just taught at a specific level (and have not been a compulsory topic when the teachers were students at the Gymnasium), this response might be due to the fact that the teachers have not yet taught on all levels. This makes it also evident that the number of teachers that participated in the evaluation appears too little to draw general conclusions. Yet, when looking more detailed at the individual materials, things are slightly different. This becomes already obvious when analysing the responses to the materials on Volta's 'crown of cups': These items were evaluated with five point Likert items, 1 meaning 'strongly agree', 2 meaning 'agree', 3 'neither agree nor disagree', 4 'disagree' and 5 'strongly disagree'. (Diagram 2):

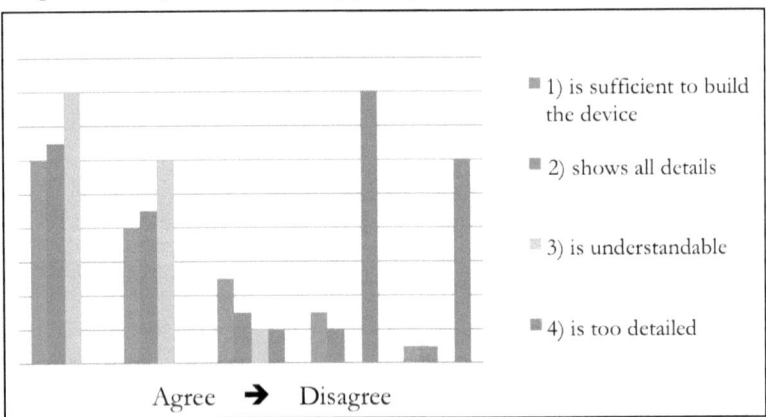

Diagram 2: Evaluation of the construction manual of the 'Crown of Cups'

Evidently, the visual manual appears to be understandable to the teachers (which is not too surprising as we considered it as extremely detailed, maybe even too detailed). However, this was seen different by the teachers, they strongly disagreed with the respective statement. Obviously, making such descriptions 'as detailed as possible' is perceived as adequate by the teachers. Yet, with respect to this material, we did not just evaluate the visual part but also the written introduction.

From the teachers responses (Diagram 3), it gets evident that the information given in this part was considered to be highly relevant (strong objection to 'superfluous') and motivating. This is also relevant as it contains part of the historical context. This can at least be seen as an indication that these teachers were not just interested in the apparatus as a decontextualised physical device but that it mattered to them to have also the historical context.

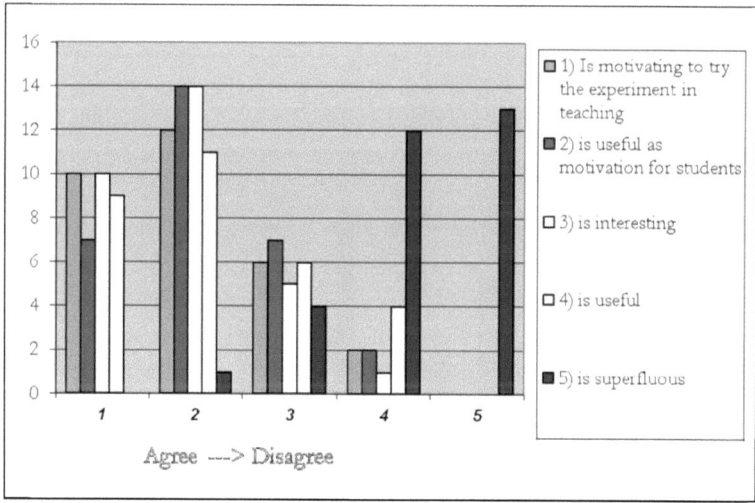

Diagram 3: Evaluation of the textual materials for the 'Cups of Crown'

Similar aspects could be seen in the responses to the early experiments of Ampère with suspended magnets, spirals, and coils. Here (Diagram 4), it gets evident that the materials are also considered as being interesting and also to be relevant to their teaching.

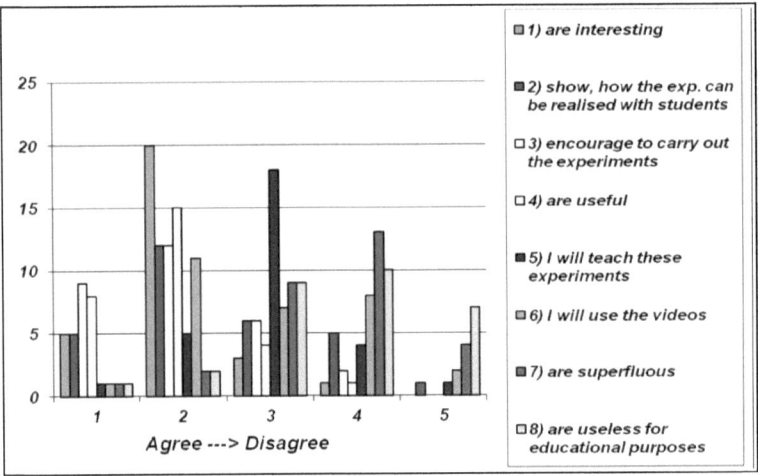

Diagram 4: Evaluation of materials on Ampère's early experiments

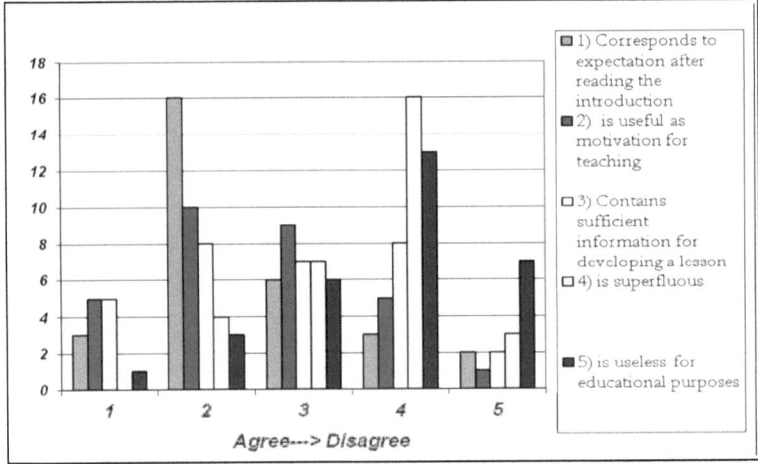

Diagram 5: Ampère's current balance video

However, it is striking that the teachers were clearly indifferent about whether to use these materials in their future teaching or not. This should be taken as an indication that some further improvement might be necessary.

Even though the teachers clearly indicate that the materials are neither useless for educational purposes not superfluous, the strong indifference with respect to the individual use in their classroom should be taken seriously. It may be an indication that here some additional materials are necessary, particularly as these materials appear to be attractive enough to not reject them for the use in educational situations.[16]

The videos on Ampère's current balance (Diagram 5) appears to be useful according to the teachers, however, at the same time it gets evident that the written materials need some modification. The responses to the video seem to justify its conceptual design, it is clearly neither perceived as 'superfluous' nor as 'useless for educational purposes'. However, things appear to be different with respect to the additional (written) materials: Obviously, some modifications are necessary as the video does not meet the expectations after reading the introduction. Unfortunately it remains unclear in which respect the expectations were not met, here, some additional research appears to be necessary. Moreover, with respect to the relevance for educational purposes, some improvements seem also to be necessary: Whilst on the one hand, a majority of the teachers indicate that the video is not useless for educational purposes; they are at the same time indifferent to the motivational aspect. Moreover, the materials seem to lack sufficient information for a significant number of teachers in order to be able to develop a lesson in which the video can be used.

At least some of the findings formulated with respect to Ampère's current balance can also be seen in the evaluation of the videos on Faraday's rotating apparatus (Diagram 6) and Jacobi's motor (Diagram 7). The videos themselves seem to be attractive as well as potentially useful to the teachers. Nevertheless, according to the responses of the teachers, neither are the initial expectations met, nor are the materials adequate for preparing a lesson in which the video can be included. This appears to be in particular a problem with respect to the Jacobi motor, however, as already pointed out, here some more information about the relevance of this material with respect to the compulsory teaching content might change the perception. Moreover, this might also be different if teachers with significantly more experience would be evaluated.

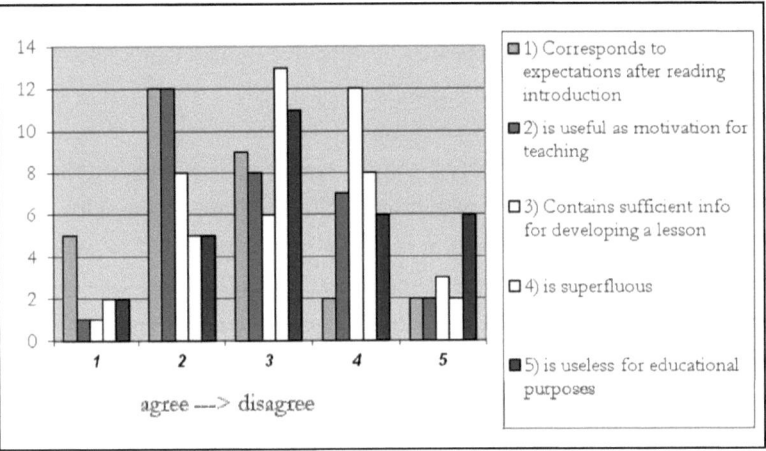

Diagram 6: Evaluation of the video on Faraday's rotating apparatus

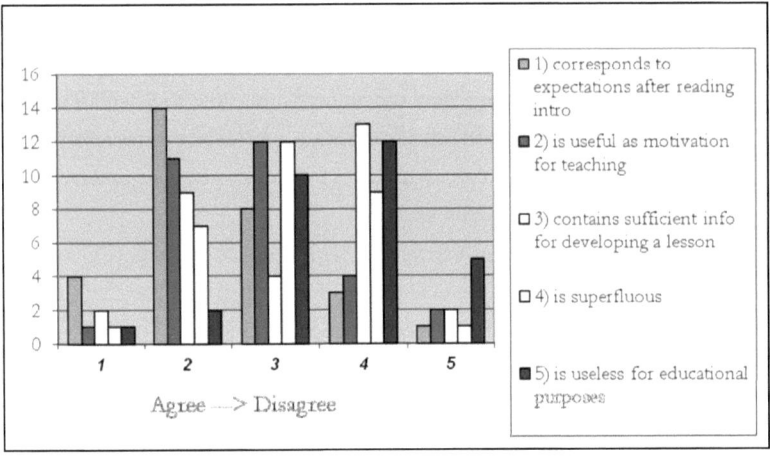

Diagram 7: Evaluation of the video on Jacobi's motor

The main deficit most teachers mentioned was the limitation of the materials to electrodynamics. Together with the evaluation of the individual materials, this can of course be read as an indication that materials of this kind appear to be useful. Additionally, and this is of particular interest, most teacher indicated that they wanted more material on the historical background of the experiments. This can be taken as an indication that these teachers were interested either to use these materials and the historical context in their teaching, or the materials made them at least interested in the historical context.

However, one should be cautious not to overestimate the results of this empirical evaluation. The number of participants is by far too small to lead to reliable results – even though some feedback is created through this evaluation, it can be taken only as an indication and in some cases inspiration to future projects in a similar manner.

Conclusion and Perspectives

We aimed at developing visual materials that can be used in the context of science teacher e-training in the field of electrodynamics. With respect to the initial intention, it has to remain an open question whether many teachers and students will use the materials. However, some of the conceptual ideas on which these materials were based appear to be justified. Particularly the distinction between those experiments which might be carried out by students in the classroom and those which can only be demonstrated appears to be quite useful for future projects.

Likewise, we feel that the idea to use materials without spoken explanations is worth considering for future projects. Initially, the basis for this decision had been to develop materials that can immediately be used by all partners of the project, despite potential language problems. Yet, during the project, another aspect became evident: Omitting any spoken or written text in the videos also enabled teachers to adapt the materials to their individual needs more easily. An open problem in this respect was the question of whether to have some background music or not. Whilst several teachers responded that this is irritating and even disturbing, the majority rated the music as being inconspicuous. However, it is not in every case possible to simply turn off the sound as some information (such as the closing of the electrical circuit) is communicated by an acoustical signal.

Surprisingly, the evaluated teachers rated the detailed account of how to build a 'crown of cups' very positive. This might be due to their lack of experiences: Being offered materials which appear to be 'fool proof' might seem to be attractive to them. Here, some evaluation with experienced teachers probably would provide additional useful information. Yet, what appears also interesting in this context is the evaluation of the materials on Ampère's early experiments. Here, the response was also positive, this can be seen as an indication that materials which are intended to support teachers' preparation of students' activities were accepted.

Necessities seem to exist particularly in two areas: on the one hand, several teachers indicated that they consider the materials inadequate to prepare a lesson in using it.[17] And they indicated that the historical background information was not as detailed as they would prefer. Here, some additional materials need to be developed, potentially in a manner where different levels of information can be offered to teachers with different backgrounds.

To summarize it can be said that visual materials which are propagated through the internet appear to be a valuable addition to the already existing materials which are intended to enable teachers to teach physics through its history. It was not a question of this study whether science should be taught through its history – The value of such an addition seems to be beyond doubt. However, what appeared to be an open question is how materials and in particulars experiments can be made to enter the classroom. Here, IT appears to have significant potential to enable teachers to have the individual perspective of being able to teach science through its history.

References

Achilles, M. 1996. Historische Versuche der Physik: Funktionsfähig nachgebaut. Frankfurt/Main: Wötzel.

Blondel, C. (1982). *A.-M. Ampère et la création de l'électrodynamique, 1820-1827*. Mémoires de la Section des sciences, 10. Paris: Bibliothèque nationale.

Brain, R. M. (2007). Hans Christian Ørsted and the romantic legacy in science: Ideas, disciplines, practices. Dordrecht: Springer.

Cavicchi, E. (2003). Experiences with the magnetism of conducting loops: Historical instruments, experimental replications, and productive confusions. *In: American Journal of Physics 71 (2)*: 156-167.

Habben, D. (1991). Experimentelle Untersuchungen zur Umwandlung von elektrischer in mechanische Energie am Modell eines frühen Elektromotors nach Jacobi. Univ. Oldenburg, teacher thesis.

Heering, P. (2011). Tools for Investigation, Tools for Instruction: Potential Transformations of Instruments in the Transfer from Research to Teaching. Experiments and Instruments in the History of Science Teaching. In: Heering, P. & Wittje, R. (eds.), *Learning by Doing: Experiments and Instruments in the History of Science Teaching*. Stuttgart: Franz Steiner, 15 – 30.

Heering, P. (2009). "From Ørsted to Jacobi - some aspects from the history of electromagnetism as background information for educators." in Kokkotas, P. and Bevilacqua, F. (eds.), Professional Development of Science Teachers: Teaching Science using case studies from the History of Science, 83-110.

Heering, P. (2007). "Educating and Entertaining: Using Enlightenment Experiments for Teacher Training." in Heering, P. and Osewold, D. (eds.), *Constructing Scientific Understanding through Contextual Teaching*, Berlin: Frank & Timme, 65 - 81.

Heering, P. (2007a): "Coils, Currents and Forces: Ampère's Experiments on Electromagnetism". In: Online Proceedings of the Ninth International History, Philosophy, Sociology & Science Teaching Conference, http://www.ucalgary.ca/ihpst07/proceedings/IHPST07%20papers/222%20Heering.pdf

Höttecke, D. (2000). How and What Can We Learn from Replicating Historical Experiments? A Case Study. Science and Education. 9 (4), 343-62.

Jungnickel, C., & McCormmach, R. (1986). *Intellectual mastery of nature: Theoretical physics from Ohm to Einstein*. Chicago: University of Chicago Press.

Kipnis, N. (2005): Chance in Science: The Discovery of Electromagnetism by H.C. Oersted, in: Science & Education 14, 1-28.

Kipnis, N. (2001). Scientific Controversies in Teaching Science: the Case of Volta. In: F. Bevilacqua, E. Giannetto, M. Matthews (Eds.): *Science Education and Culture: the Contribution of History and Philosophy of Science*. Dordrecht: Kluwer, 255-271.

Kipnis, N. (1993). *Rediscovering Optics*. Minneapolis: Bena Press.

Kokkotas, P. & Bevilacqua, F. (eds.) (2009). *Professional Development of Science Teachers: Teaching Science using case studies from the History of Science*

Leister, H. (2008). Evaluation der in Oldenburg im Rahmen des Projects Science Teacher e-training entwickelten webbasierten Materialien. Univ. Oldenburg, BA-Thesis.

Lühr, J. (2000). Die Geschichte eines Demonstrationsexperiments: Zur Geschichte der Ampère'schen Stromwaage. In P. Heering; F. Rieß; C. Sichau (Eds.): Im Labor der Physikgeschichte: Zur Untersuchung historischer Experimentalpraxis. Oldenburg: BIS, 135-156.

Martins, R. d. A. (2003): Resistance to the discovery of electromagnetism: Ørsted and the symmetry of the magnetic field, in: F. Bevilacqua & E. Giannetto (Eds.): Volta and the history of electricity. Pavia / Milano: Università degli Studi di Pavia /Editore Ulrico Hoepli, 245-265.

Meya, J. (1990). Elektrodynamik im 19. Jahrhundert: Rekonstruktion ihrer Entwicklung als Konzept einer redlichen Vermittlung. Wiesbaden: DUV.

Oersted, H.C. (1820). Experimenta circa effectum conflictus electrici in acum magneticam. In: Journal für Chemie und Physik 29, 275-281.

Reinhold, P. (2000). Open Experimenting – An Approach to Structure Science Teaching and Learning. In: S.Hopmann; K. Riquarts; I. Westbury (eds.) *Teaching as a Reflective Practice. The German Didaktik Tradition*, Erlbaum, Mahwah (New Jersey), 295-318.

Sibum, H.O. (2006). Maschinen, Fledermäuse und Schriftgelehrte: Experimentalwissen im späten 18. und 19. Jahrhundert. In: H. Schramm; L. Schwarte; J. Lazardzig (Eds.): *Instrumente in Kunst und Wissenschaft: Zur Architektonik kultureller Grenzen im 17. Jahrhundert*. Berlin: Walter de Gruyter, 302-318

Steinle, F. (2005). Explorative Experimente: Ampère, Faraday und die Ursprünge der Elektrodynamik. Stuttgart: Steiner.

Teichmann, J. (1979). "¬Die Rekonstruktion historischer Modelle und Experimente für den Unterricht - drei Beispiele." *In: Physik und Didaktik Band 4*: S. 267 - 282.

Volta, A. (1800). On the Electricity excited by the mere Contact of conducting Substances of different Kinds, in: The Philosophical Magazine 7, 289-311, (reprinted in A. Volta, On the Electricity excited by the mere Contact of conducting Substances of different kinds: Bicentenary Edition in French, English, German and Italian of the Letter to Sir Joseph Banks of the 20th of March 1800. Milano: Ulrico Hoepli, 1999, 31-53.

[1] This project was financially supported by the EU Socrates programme, programme number 129552-CP-1-2006-1-GR-COMENIUS-C21

[2] The partners in the project were: U. Athen (P. Kokkotas), U. Thessaloniki (F. Seroglou), U. Zypern (N. Valanides), U. Pavia (F. Bevilacqua), U Oldenburg (P. Heering). For some outcomes see Kokkotas & Bevilacqua 2009, in which some ideas discussed in this paper are already published. Moreover, the materials are available at the webpage http://stet.wetpaint.com.

[3] In this respect one could also discuss the relation between research experiments and teaching demonstrations that are based on this teaching experiment (Heering 2011).

[4] To give but one example where such difficulties might be identified: Electrical experiments from the Enlightenment are extremely well suited for teaching purposes as they were performed as public demonstrations and central concepts of electricity were developed in this period (Heering 2007). However, in most schools the necessary static frictional generator is not to be found. Instead, there are Wimshurst machine's which are more powerful (and thus less useful for making discharge experiments where the human body is involved) and conceptually much more complex. As a result, its working principle remains unclear to the students.

[5] This is of course not an adequate discussion of these experiments; however, this is also not the focus of this paper. There are numerous publications on the history of early electrodynamics, for a more thorough discussion of these experiments in the context of the STeT-project see Heering (2009).

[6] These materials can be found at http://stet.wetpaint.com/page/Galvani%27s+Experiments, last access March 8th, 2011. These simulations were developed by the Pavia group led by F. Bevilacqua.

[7] Apart from preparing visual materials on the devices and experiments of both researchers, we also prepared materials on the controversy between Volta and Galvani on the conceptual explanation of these experiments.

[8] The materials can be found at http://hidistet.wetpaint.com/page/Voltas+Tassenkrone, last access March 8th, 2011.

[9] However, immediately after Oersted's publication, several researchers published claims that they had observed a magnetic action of an electric current earlier. Even though at least skepticism with respect to these claims is necessary, they illustrate that there was a strong belief that a connection between electrical and magnetic actions exists. This was not limited to galvanic electricity, already in the 18th century the quest for a relation between static electricity and magnetism was an issue.

[10] The video demonstrating the operation of the current balance can be found at http://www.youtube.com/watch?v=_qw5FHjmZY8, last access March 6th, 2011.

[11] In this respect, exploratory experiments show remarkable similarities to the concept of 'open experimentation' (Reinhold 2000), for a more thorough discussion of Ampère's early experiments on electromagnetism see Heering (2007a).

[12] The videos can be found at http://hidistet.wetpaint.com/page/Amperes+Vorversuche, last access March 6th, 2011.

[13] On Faraday's rotating apparatus as well as on experiences made in reconstructing this device and educational version that materialize the working principle see Höttecke (2000). Fig 6 shows the reconstructed device as it was demonstrated in the video, for the

video see http://www.youtube.com/user/histodid#p/u/22/kTOVferbAco, last access March 6th, 2011.

[14] On the reconstruction of the Jacobi motor see Habben (1991), on a historical discussion Sibum (2006).

[15] The video can be found at
http://www.youtube.com/user/histodid#p/u/21/8TaeJGDJlcA, last access March 6th, 2011.

[16] The questions to the materials on the 'crown of cups' and the early experiments by Ampère were somewhat different than those on Ampère's current balance, Faraday's rotational apparatus and the Jabobi motor. This was related to the differences in the intended use of the visual materials: whilst the former were supposed to be the basis of students' activities, the latter were intended to be shown in the classroom.

[17] It remains open whether the teachers expect completely prepared lessons, however, this seems to be at least with respect to German teachers doubtful.

Using educational ICT to include history in science teaching and in science teacher training

Pere Grapí

Centre d'Estudis d'Història de la Ciència (CEHIC), Universitat Autònoma de Barcelona, Spain.

ABSTRACT: The growing relevance that the understanding of the nature of science is acquiring in secondary science education has encouraged some researchers to look for new learning pathways to help students to understand the more salient issues of the nature of science. This article deals with this prospect from two different and complementary approaches. Primarily, a presentation of how certain ICT learning activities – from the simplest to the most refined – based on the history of science, can become powerful resources to arrive at an appropriate knowledge of the nature of science. However, if science teachers remain unaware of the history of the science they teach, this latter aspect may remain unattainable, and therefore the second part of the article is devoted to presenting an online pilot course of history of science for science teacher training in order to bridge the instructional gap of science teachers in this subject.

Presentation

The purpose of this article is twofold: The first part deals with the potentiality of certain ICT resources for including history in science teaching, while the second part is devoted to presenting an online course of history of science for in-service science teachers that was launched in Catalonia during the academic year 2009-2010.

Any discussion about ICT resources for an immersion of the history of science and technology (HST hereafter) in science teaching must inevitably involve the growing impact of the "digital revolution" on academic research about the history of science in recent years. The journal *Isis* for the year 2007 published two articles under the generic heading of *Clio Electric* that provided some initial reflections on this issue. The first of these articles by Robert Hatch initiated a discussion about those digital resources that the internet could provide. In this regard, the author suggested some topics to be addressed, for example: textual, graphical, visual or oral documentation, digital publications and virtual museums (Hatch 2007). The second article by Bryan Dolan, while taking up the challenge of the first article, discussed the need for replacing academic communication and research within a new agenda for a more democratic access to information (see Annex 1 for internet sources of digitized documents on the history of science), all with the final aim of recasting the cultural heritage of humankind, by digitizing libraries and museums in order to make them accessible through internet (Dolan 2007)[1].

Yet quite apart from the academic interest in the arrival of the internet in the field of the history of science, certain internet resources have also been making their own way in the university curricula[2]. Nowadays it is possible to find readings, forums, multimedia animations, films, images, etc. belonging to the field of the history of science as part of the contents of university courses. The unavoidable reality is that in universities as well as in secondary education, the internet is becoming a permanent and customary place of learning. This is a reality that we need to accept as soon as possible rather than dismiss, in spite of all the challenges it presents and the disadvantages that also need to be taken into account. (Gooday 2003; Sumner 2003). The history of science now has the chance to play its own role in this new situation, and with a much greater probability of success than in the past. Meanwhile, in recent years, the internet has gradually been gaining ground as a medium for sharing knowledge, relegating its function of a mere content provider to a secondary role. The outgrowing social nets are becoming the windows of this new internet paradigm known as Web 2.0 or Web 3.0

The nature of science, the history of science and science education.

In the field of science education a substantial agreement exists that both scientific instruction and science teacher training involve something more than an understanding of the fundamental facts and explanations in the domain of normal science; it also requires knowledge about science itself, about how scientific knowledge has been obtained, how reliable it therefore is, what its limitations are, and to what extent we can rely on it and its changing methods, as well as about the interface between scientific knowledge and the wider society. In brief, to know something about the nature of science itself[3].

On conclusion of their secondary education, students should have acquired skills on the nature of science such as being able to know and understand aspects of scientific explanations, scientific methodology, the range of explanations of science, scientific creativity, the role of the scientific community and the implications of science in their lives. These skills might be acquired by developing the conceptual and procedural contents concerning "normal science" in the sense that Thomas Kuhn gave to this term. Alternatively, however, general topics on the nature of science can also be achieved by placing scientific knowledge in its own historical and philosophical context.

Specifically, a student should be able to know and understand that: (1) good scientific explanations lead to predictions; (2) there is no simple scientific method to provide knowledge automatically, although the work of scientists has some differential and characteristic trends; (3) different kinds of explanations exist, among which there are hypotheses, laws, theories and models proposed to explain data; (4) the making of explanations is a creative process that depends not only on experimental data, but also on the experimenter's mental scheme and his or her cultural and social context, (5) the scientific community has established procedures to contrast the discoveries and conclusions of scientists with the aim of reaching agreements and consensus, and (6) the application of scientific knowledge to new technologies, materials and appliances has an impact on people's lives, especially as regards their unexpected or unwanted side effects. The following sections are devoted to displaying the usefulness of certain ICT resources for both science education and science teacher training; the learning activities associated with these resources are precisely linked to some aspects of the nature of science.

The management of certain ICT learning activities based on the history of science.

This section is devoted to presenting several learning activities aimed at making the presence of history available to students while learning some science content. These activities are designed to be incorporated into a course created using the popular LMS (Learning Management System) - known as Moodle (Module Object-Oriented Dynamic Learning Environment) - an educational platform based on social constructivism. Educational social constructivism argues that a person builds his knowledge by sharing and contrasting his ideas with the ideas of the others - experts and peers – and from this collaborative process the person is able to crystallize his own knowledge.

The internet is increasingly becoming a channel for the exchange of information, and the Moodle platform is a good example of how the internet can contribute to sharing information for educational purposes. Moreover, since the epistemological background of Moodle is social constructivism, this platform tends to reinforce all those activities aimed at collaborative learning. Furthermore, due to the fact that the HST provides episodes that make the existence of disputes and debates in science evident, then the coupling of these two prospects contributes to producing learning activities in which students share their knowledge while dealing with the historical context of certain scientific events. Moodle modules such as forums, chats, wikis and glossaries enable these types of activities to be created.

Forums and glossaries

With regard to the topic of air pressure, the following forum has been proposed to secondary school students aged 14-15. Its starting point is a well known experiment that consists in filling a glass with water, covering it with a piece of paper, turning it upside down and observing how water does not leak out of the glass (Fig. 1).

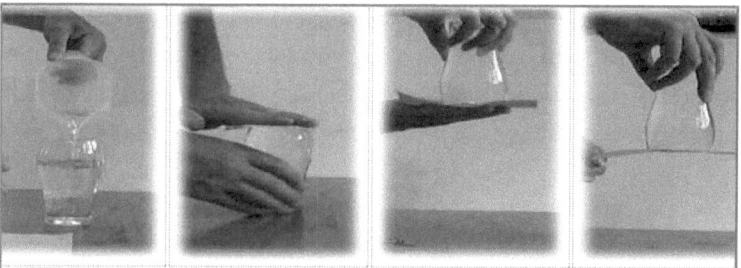

Fig. 1: Steps to carry out the experiment under discussion in the forum

Atmospheric air is an ideal subject for making the history of science perceptible in science education. The fact that air is a gaseous mixture characterized by its optical transparency has contributed to its becoming the kind of material that people often neglect as far as its physical characteristics and its role in many chemical phenomena. Needless to say, the Aristotelian conception of the air as one of the four natural elements lasted until the end of the eighteenth century.

The historical background of the activity is based on nature's abhorrence of a void according to the ideas of Aristotle and his followers, and their corresponding belief in the weightlessness of air. So for the Greek philosopher and his many followers, air had no weight exerted no pressure. This position was developed in the Middle Ages through the theory of "horror vacuity", according to which Nature's abhorrence of a void led her to employ all the means at her disposal to prevent its occurrence. In the seventeenth century, the debate on the void and atmospheric pressure represents one of the key points in the discussion about the constitution of matter and the nature of the universe. This discussion reached its peak when Evangelista Torricelli (1608-1647) challenged the idea of the "horror vacuity" after his famous experiment conducted in Florence in 1644 Students are requested to explain the phenomenon presented in that forum by arguing from both the

Aristotelian idea of nature's horror of a void and the ideas of Torricelli and Pascal, grounded on the existence of atmospheric pressure. This is an experiment that could be done as easily two thousand years ago as today. The participants in this forum should strive to give their explanations from different scientific perspectives, realizing that both perspectives can provide theoretically coherent explanations. Students can also contradict, modify or support each other's explanations; ultimately, they can participate online in a scientific controversy, and in doing so they are learning a characteristic trend in the nature of science.

Fig. 2: Detail from the frontispiece of Riccioli's *Almagestum Novum*

The following forum was designed for an online course of history of science for in-service science teachers. The focus of this forum was the picture below (Fig. 2), which shows an allegory about the world-systems of

Ptolemy, Copernicus and Tycho Brahe. This is an illustration from the book *Almagestum Novum,* written in 1651 by Ginvonni Battista Riccioli. Participants in the forum were requested to identify the location of these three world-systems and subsequently to discuss Riccioli's assessment of the three systems according to their relative position in the picture. The activity proved to be quite satisfactory, since as frequently happens when analyzing inscriptions such as maps, tables, pictures, lists, etc., people discover different elements in the inscription that help to complement their explanations. What follows are two selected fragments of the entries in this forum:

Gemma

I think that the system represented at the bottom is Ptolemy's; the system represented in the lower plate of the balance is Brahe's, and finally, the system represented in the upper plate is Copernicus'. Apparently, it seems that the system placed at ground level has been rejected by the goddess of justice. According to the inclination of the balance, the most consistent system would be Brahe's. If so, the author valued each newly proposed system as better than the previous one, i.e, as if advances in knowledge followed a chronological order. It could also be that Ptolemy's system was considered too good to be compared with the other two and therefore it was out of discussion. This would be a more appropriate position for a Catholic priest such as Riccioli.

According to his biography in Wikipedia, Riccioli was a Jesuit priest who opposed the Copernican system because it was contrary to the biblical narration of the place of the Earth in the universe. However, it seems that although Riccioli valued it as a mere hypothesis, he manifested a certain sympathy for Copernicus since he named a prominent Moon crater after Copernicus.

Manel

[…] I find that the figure on the left is Argos, holding a telescope to observe the Sun. At the bottom is Ptolemy himself with his own system on the ground. Ptolemy leans his hand against the coat of arms of the Prince of Monaco, to whom the "Almagestum Novum" was dedicated. At the top there are the century's newest astronomical discoveries: Mercury, and Venus showing its phases; Saturn is still pictured without its rings, Jupiter with its four moons and two parallel bands on its equator, and something that looks like a meteorite moving like a cannonball.

Another type of collaborative activity consists in making a glossary from the biographies of outstanding protagonists in different science subjects. Unlike forums, the aim of glossaries is not so much contrasting points of view but to enhance individual contributions for the benefit of all the participating students. The biographies of scientists who emerge during the course usually capture the attention of the students - the life of the others generally tends to arouse interest. The activity aims to focus this interest onto protagonists of science by making them more familiar to the students and sometimes to demystify them. Fig. 3 shows an entry for a glossary of scientific biographies made by a 14 year-old student.

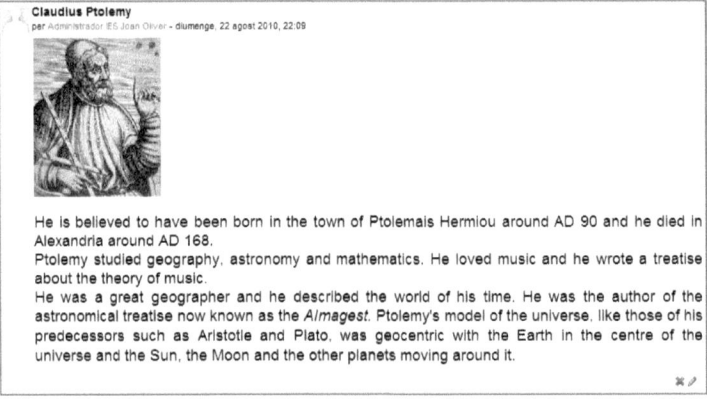

Fig. 3: Screenshot of the entry "Ptolemy" for a glossary of scientific biographies

Questionnaires

As regards the presence of the history of science in science teaching, it is important to recognize the role that the internet is developing as a library of contents, such as the audiovisual materials that provide virtual replications of historical experiments. Suffice it to say that nowadays the websites of some museums that keep collections of scientific and technological material have become good multimedia locations for the history of science. In fact, all those museums that preserve the material culture of science are in a unique position to produce virtual reconstructions of the past that we wish to teach. These resources have the virtue of combining narration with interpretive images of objects, contextualized sceneries, animations and even original auditions (Borda & Bud 2003). In short, these resources are able to offer virtual reconstructions of the past that the print media can hardly achieve. (See Annex 2 for websites of some museums providing multimedia resources of history of science)

These virtual replications can for instance be used as an introductory part of questionnaires. On the one hand, animations allow for the virtual replication of experiments, recreating their historic scenery and providing students with a vision of the experiment in its historical context. On the other hand, Moodle allows for the creation of learning activities – e.g., questionnaires – by embedding online animations.

Coming back to the subject of atmospheric pressure, the *Museo Galileo*[4] in Florence hosts excellent virtual replications of benchmark experiments such as those by Torricelli, Pascal and von Guericke. A screenshot of Torricelli's experiment is shown in Fig, 4.

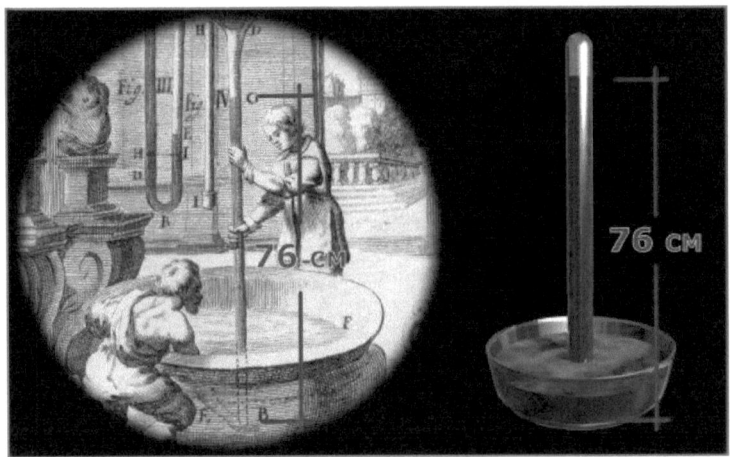

Fig. 4: Screenshot of the virtual replication of Torricelli's experiment

Fig. 5: Screenshot of a virtual reconstruction of Pascal's experiment

Torricelli stated that people lived at the bottom of an ocean of air and that air at the summit of a mountain weighed less than at its foot. Blaise Pascal took up this idea and predicted that if Torricelli was right then the amount of mercury inside a tube containing atmospheric air should be smaller at the summit of a mountain than at its foot. In 1648, Pascal instructed his brother-in-law to climb to the summit of the Puy de Dôme (1400 m) to corroborate this prediction. For this case, a virtual reconstruction of a similar experiment carried out by academicians of the *Accademia dei Cimento* is shown in Fig. 5.

Another virtual replication of experiments concerning the effects of air pressure is that carried out by Otto von Guericke in 1656 in Magdeburg (Fig. 6). Von Guericke used two perfectly matching hollow bronze half-spheres that he fastened together tightly to form a sphere. Using an air pump, he extracted the air from this sphere. Once the vacuum had been created he removed the external connectors between the two hemispheres and was able to show the astonished spectators how two teams of eight horses pulling in opposite directions barely sufficed to separate the two emptied hemispheres. This experiment demonstrated the force of atmospheric pressure when it is not offset by the internal pressure of the hemispheres.

Fig. 6: Screenshot of the virtual replication of von Guericke's experiment

Apart from Moodle learning activities, we need to benefit from those good applications that exist on the internet and can be of help to engage students. One of these applications is a timeline, which consists in placing characters chronologically and adding their biographies. The timeline can also allow events and scientists to be added in their social, artistic, literary, religious and philosophical context (Fig. 7).

Fig.7: Screenshot of a timeline of scientists

The other activity is geomapping, which consists in geographically locating different scientists to acquire a more local view of their scientific activities (Fig 8). Needless to say, entries in the glossaries, events in the timeline and the locations on the map are added by the students themselves rather than by the teacher.

Fig. 8: Screenshot of a map with geographical locations of scientists

The need for instruction in HST for in-service science teachers

In the field of science education there exists a substantial agreement about the need for implementing the history of science in both scientific instruction and science teacher training. This is not only necessary but possible, since we now have more accessible and engaging resources to achieve this than, for example, ten years ago, as outlined above. However, a further step forward is needed to make this possibility a reality.

The following anecdote reveals a major constraint in this enterprise. In the year 2005, the Education Section Committee of the BSHS decided to run a competition for teachers based on the History of Radioactivity. The committee decided to give a prize of £500 (€750) for three one-hour lessons on the History of Radioactivity. The competition was advertised at conferences and in regional and national publications. The result was that they received no (zero) entries despite this extensive publicity (Fowler 2008, 771). This anecdote illustrates one of the reasons that among others are usually mentioned when referring to the relative absence of the history of science in science teaching: the supposed reluctance of science teachers to place HST in their science lessons and their apparent low level of interest in HST (Rudge 2007, 238). These are two reasons that feedback on themselves, forming a vicious circle that is very difficult to break. However, teachers can only teach what they know, and in order to be convinced of the usefulness of the history of science is absolutely essential for them to fill their training gap in this subject. The key is to transform a vicious circle into a virtuous one.

In order to convince teachers of the usefulness of HST for science learning, they must fill their instructional gap in that subject to enable them to appreciate the learning values of HST. The training of in-service science teachers in HST shows some parallelism with the teaching of HST to undergraduate science students (Cantor 2007; Gooday 2007), with the important difference that the former have gained some teaching experience, which provides them with a more complete understanding of the science they are teaching.

Ordinary in-service science teachers are not expected to have enrolled as undergraduate students in a historical course during their scientific career; they have been instructed in a particular science understood as normal science. In general, in order to progress in science, science teachers acquire learning strategies that may be an inconvenience when they take a course in HST. In this regard, in-service science teachers may have to overcome substantial difficulties that should be kept in mind when developing a training course in HST. By way of example, they will have to tackle issues such as: (1) understanding the value of the debate between opposing or complementary theories, (2) accepting the ordinary disagreement between different HST sources, (3) designing possible answers for a controversial situation instead of looking for the right answer, (4) avoiding the tendency to "whiggish" stories that often have a modern correct ending, (5) improving their knowledge of the past historical context of the events they have to deal with, (6) working simultaneously with different documents to extract relevant information, or (7) assessing historical accounts composed by their colleagues. Needless to say, some salient features of the nature of science can readily be recognized in this list.

The coming of an online HST course for in-service science and mathematics teachers

The educational background of the course

In the academic year 2007-2008, a new educational curriculum for secondary education started in Catalonia. One of the main differences as regards the previous science curriculum is that the new one put the nature of science at the heart of one of its general aims. The tenth general aim of the science curriculum states that science learning should develop the capacity to recognize the nature of science and place the most relevant scientific knowledge in a historical context, in order to understand the genesis of fundamental concepts and theories of science as well as the interactions among science, technology and society. A similar situation occurs in the new curriculum for mathematics, where some notions of the historical genesis of salient topics are included as part of the general content of the subject.

Unfortunately, however, in education there is often an enormous distance between a rule and its accomplishment. For instance, it is important to remember that the introduction of the National Curriculum for Science in the UK in 1989 included an attainment target (AT17) devoted to the nature of science, which caused some alarm among teachers, and when it disappeared in the revision of the National Curriculum in 1992 (Ellis 2005) they breathed a sigh of relief.

In Catalonia during the 80's and 90's, a significant number of science teachers were committed to the history of science when universities began to offer postgraduate degrees in the history of science. At that time it was relatively easy for in-service teachers to enroll as part time students in these courses, which carried a Master or PhD degree in the history of science. However, important and abrupt changes in the professional conditions of secondary school teachers in the last ten years, together with the increasing costs of postgraduate course fees, have led to a shortage of science teachers taking these courses.

Nowadays, the history of science has no securely established role in the university science curriculum. The history of a particular science such as physics, chemistry, biology, mathematics or technology is a matter that is left to students themselves to decide. Up to the academic year 2008-2009, university students in their last year who decided to take an optional course in order to gain a teacher-training certificate qualifying them for teaching in secondary education might have taken one or two sessions devoted to the history of a particular science[5].

A new opportunity for HST training has arisen in recent years with the arrival of the new online training courses for in-service teachers sponsored by the same Department of Education in Catalonia. In order to enable science teachers to use the history of science in their science courses, and also to be trained in the history of the science they are teaching, the Catalan Society for the History of Science and Technology (SCHT) produced an online HST pilot course for in-service teachers for the Centre of Documentation and Experimentation in Sciences (a service of the Department of Education). This is one of the few courses for online learning in HST. Nevertheless, it is worth pointing out that for Spanish speaking teachers the National University of Distance Education (UNED) offers two semester courses on the history of chemistry in its programme for in-service teacher training. For the English speaking world, the prestigious Open University (OU) has a variety of courses in the History of Science, Technology and Medicine as an undergraduate subject in Arts and Humanities, Health and Social Care, and Science. The same Open University also offers several free learning units related to the history of science (Open Learn). For those teachers looking for an introduction to what the internet has to offer and how to retrieve the information they need on the history and philosophy of science, there is the free and interactive tutorial "Internet for the History and Philosophy of Science" guided by a team from Leeds University (See Annex 3 for the URLs of these online courses)

Presentation of the course

The term "online course" should be understood as an approach to both distance and open learning. It is "distant" in the sense that a significant proportion of the teaching is conducted by someone removed in space and/or time from the learner, although some sessions of the course need to be conducted face-to-face, and it is "open" because it is a course in which constraints on study are minimised in terms either of access, time, place, rhythm or method of study. The fact that participants are able to self manage their own learning process has proved to be a decisive factor in the well-known success of this sort of learning. Actually, the term "online" evokes a course managed through internet where computer-based technologies can be used for two-way or multi-way communication.

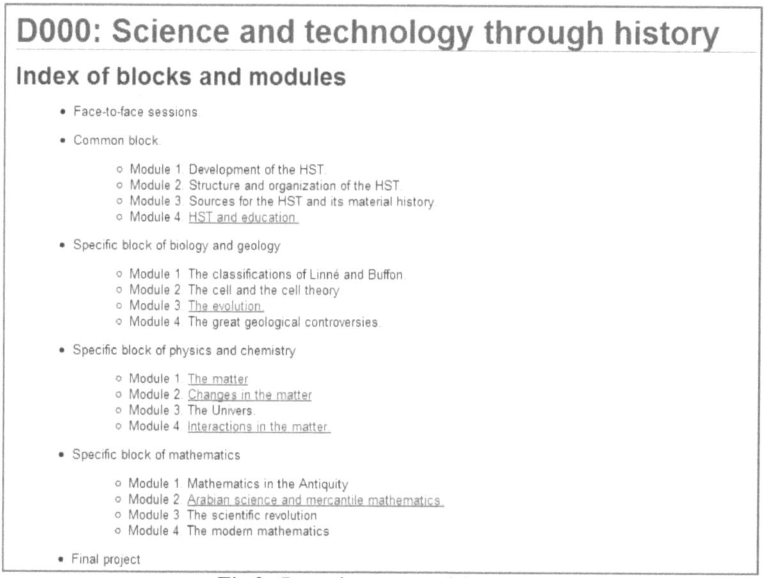

> **D000: Science and technology through history**
> **Index of blocks and modules**
> - Face-to-face sessions
> - Common block
>   - Module 1. Development of the HST
>   - Module 2. Structure and organization of the HST
>   - Module 3. Sources for the HST and its material history
>   - Module 4. HST and education
> - Specific block of biology and geology
>   - Module 1. The classifications of Linné and Buffon
>   - Module 2. The cell and the cell theory
>   - Module 3. The evolution
>   - Module 4. The great geological controversies
> - Specific block of physics and chemistry
>   - Module 1. The matter
>   - Module 2. Changes in the matter
>   - Module 3. The Univers
>   - Module 4. Interactions in the matter
> - Specific block of mathematics
>   - Module 1. Mathematics in the Antiquity
>   - Module 2. Arabian science and mercantile mathematics
>   - Module 3. The scientific revolution
>   - Module 4. The modern mathematics
> - Final project

Fig 9: General structure of the course

The name of the pilot course is "Science and Technology through History". It was launched as a pilot course in the academic year 2009-2010 and lasted for four months (from November to March). The course is divided into a pair of face-to-face sessions, a common block, three specific blocks and a final project. The common block is compulsory for all the students and its general objective is to provide elements of the historiography of science to facilitate the understanding of the subsequent specific blocks. Upon completion of this common block, students must choose one specific block according to their academic training or to their teaching practice. For this pilot course three options are provided: physics and chemistry, mathematics and, biology and geology. While students are completing their specific block they also have to complete their final project, which is intended for applying what they have learnt in the design of classroom learning activities for secondary education students. This final project is compulsory for successful completion of the course and must be submitted upon completion of the specific block (Fig. 9)[6]. The content of the course is freely available online and is managed through a virtual course with the aid of the Moodle platform. However, this Moodle companion course is only available for registered students.

The structure of the course according to the contents of the physics and chemistry block

Each block is divided into four modules. The modules are the basic units of the course and each one consists in three activities; exercises, classroom resources and further references (Fig. 10). Throughout the development of the theoretical content of each activity, students find two interleaved practicals. The purpose of these practicals is to allow students to rethink some aspects of the content of the activity. These practicals do not have to be submitted for assessment, but each one is associated with a forum in the companion Moodle course with the aim of generating multi-way communication. The following practical is included in the activity: "Chemical change and electricity: Galvani, Volta and Faraday" of the module "Changes in the matter":

Volta's pile. An episode of scientific creativity

> Why did Volta change from a battery of the "crown of cups" type to the final "pile of discs" type? What other kind of electrical phenomena were under discussion during the process of discovery of his pile? We suggest reading the final part of the letter sent by Volta in 1800 to the secretary of the Royal Society of London notifying his discovery. You are asked to discover the

key step in the creative process that probably led Volta "to pile" metal discs to build his definitive pile.

To complete each module, the students are required to do their final exercises. The purpose of the exercises is to deal with some aspects of the practicals but in a deeper and a less guided way. Unlike practicals, exercises must be submitted for assessment to validate the corresponding module. What follows is one of the exercises from the module itself:

The following <u>document</u> refers to the description of the experiment that Lavoisier conducted on the combustion of phosphorus. Lavoisier began this experiment in late 1772 and on October 20th of the same year he sent a note to the secretary of the Paris Academy of Sciences describing how phosphorus absorbed air when burning to form the acid spirit of phosphorus, and that as a result of the combustion the weight of the sample increased. The interest of the document lies both in the presentation of Lavoisier's theory of combustion and in the use of the principle of the conservation of matter.

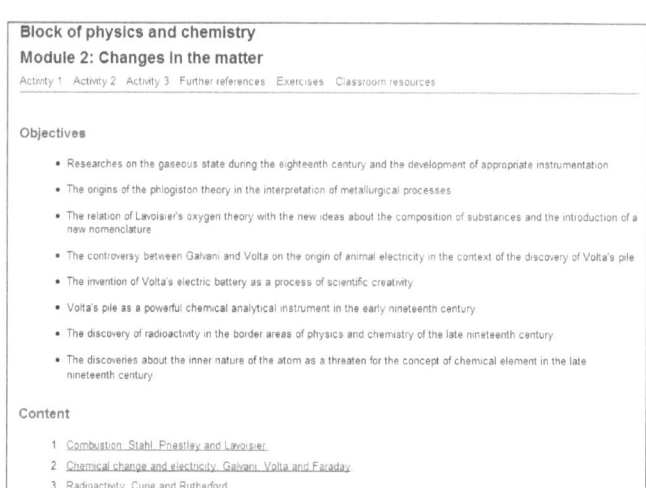

Fig. 10: Organization of a particular physics and chemistry module.

What aspects would you highlight in the planning and design of the experiment?
Identify any of Lavoisier's previous theoretical and methodological assumptions that can be detected in this description.
What conclusions did Lavoisier draw from this experiment?
Have you observed any unexpected features in the quantitative data?

Concluding remarks

The prominent role that the nature of science is destined to play in science education will require many kinds of learning resources to enable secondary school students to acquire a reasonable understanding of the main features of the nature of science. The History of Science can become an appropriate approach for constructing learning activities leading to the fulfillment of that purpose, and in this sense the use of the HST in science teaching presents a challenge for the science curriculum in secondary education.

The internet is consolidating its status as a permanent place for learning, and its accompanying ICT applications–ranging from simple animations to sophisticated virtual learning environments such as Moodle – are inexorably pervading classroom life. This article shows how the internet and educational ICT can facilitate the development of activities to boost the presence of history in the teaching of science with a greater prospect of success than it has had hitherto.

If HST is destined to be a benchmark for the teaching of the nature of science in science education, then it seems mandatory that science teachers should become acquainted with the history of the science that they teach, otherwise whatever efforts are made to make the HST visible in science learning will be rendered useless and time-consuming. The online courses in the HST available through the internet can contribute significantly to science teacher training in the HST. As coordinator of the online pilot course "Science and Technology Through History" - launched recently in my country – the first and immediate impression is that the course was completed by a significant number (more than a fifty percent) of the teachers initially enrolled, who valued it positively for their professional training. Obviously, further progress of this pilot course in the coming years is needed for a more accurate and deeper assessment.

References

Borda, A; Bud, R. (2003), 'Engaging with Science and Culture: Major Missions across Cyberspace to Share Good History'. In D. J. Moseley (ed.) *The Challenges of using the World-Wide Web in Teaching History of Science*, Philosophical and Religious Studies Subject Centre, Learning and Teaching Support Network, Leeds, pp. 26-32, http://prs.heacademy.ac.uk/publications/histscibook.pdf (accessed 28 Feb. 2011)

Cantor, G. (2007), 'Teaching Philosophy & HPS to Science Students', The Higher Education Academy. Subject Centre for Philosophical & Religious Studies, Leeds, http://prs.heacademy.ac.uk/view.html/prsdocuments/27 (accessed 28 Feb. 2011)

Dolan, B. (2007), 'Clio Electric: Online Resources in Post-1750 History of Science' Medicine, and Technology', *Isis*, 98, 2, 355-360

Ellis, P. (2005), 'Putting History in Science: Resources for Teaching History of Science Component of Science Courses in Schools in the United Kingdom'. In P. Grapí & M.R. Massa (eds.), *Actes de la I Jornada sobre la Història de la Ciència I l'Ensenyament*, IEC-SCHCT, Barcelona, pp. 9-18

Fowler, P. (2008), 'History of Science in Secondary Science Education in England'. In H. Hunger ; F. Seebacher & G. Holzer (eds.), *Styles of Thinking in Science and Technology. Proceedings of the 3rd International Conference of the European Society for the History of Science*, Vienna, pp. 771-778.

Gooday, G. (2003), 'The Challenge of Using World-Wide Web Resources to Enhance Student's Learning of History/Philosophy of Science, Technology and Medicine'. In D. J. Moseley (ed.) *The Challenges of using the World-Wide Web in Teaching History of Science*, Philosophical and Religious Studies Subject Centre, Learning and Teaching Support Network, Leeds, pp. 5-10

Gooday, G.(2007), 'The Challenges Of Teaching History & Philosophy Of Science, Technology & Medecine To Science Students', Subject Centre for Philosophical & Religious Studies, Leeds,
http://prs.heacademy.ac.uk/view.html/prsdocuments/66 (accessed 28 Feb. 2011)

Hatch, R.A. (2007), 'Clio Electric: Primary Texts and Digital Research in Pre-1750 History of Science', *Isis*, 98, 1, 150-160

McComas, W. F. (2000), 'The Principal Elements of the Nature of Science: Dispelling the Myths'. In W. F. McComas (ed.) *The Nature of Science in Science Education. Rationales and Strategies*, Kluwer Academic Press, Dordrecht, pp. 53-70.

Rocke, A. (2008), 'Letters to the Editor', *Newsletter of the History of Science Society*, 37, 2, 3

Ross, S. (2008), 'Wikipedia and the History of Science', *Newsletter of the History of Science Society*, 37, 1, 1 and 6

Rudge, D.W. (2007), 'History of Science in the Service of Middle School Science Teacher Preparation'. In P. Heering & D. Osewold (eds.) *Constructing Scientific Understanding through Contextual Teaching*, Frank & Time, Berlin. pp. 227-242.

Sumner, J. (2003), 'A Web Development Policy for History of Science, Technology and Medicine Teaching'. In D. J. Moseley (ed.) *The Challenges of using the World-Wide Web in Teaching History of Science*, Philosophical and Religious Studies Subject Centre, Learning and Teaching Support Network, Leeds, pp. 22-25, http://prs.heacademy.ac.uk/publications/histscibook.pdf (accessed 28 Feb. 2011)

Annex 1[7]

Internet sources of digitized documents on the history of science.

Gallica (Bibliothèque Nationale de France): http://gallica.bnf.fr/
Conservatoire Numérique des Arts et Métiers: http://cnum.cnam.fr/
Bibliothèque Internuniversitatire de Paris:
    http://www.bium.univ- paris5.fr/histmed/debut.htm
Bibnum: http://www.bibnum.education.fr/
The Turning Pages (The Royale Society): http://royalsociety.org/turning-the-pages/
Google Book Search: http://books.google.com/
Museo Galileo de Firenze - Biblioteca Digitale:
    http://www.museogalileo.it/en/explore/libraries/digitallibrary.html
Discórides (Universidad Complutense de Madrid):
    http://www.ucm.es/BUCM/foa/english_version/dioscorides.htm
Des ouvrages numérisés en histoire des sciences (Université de Bordeaux):
    http://cyber.bu.u-bordeaux1.fr/ouvrages/
Archives de l'Académie des Sciences de Paris:
    http://www.academie-sciences.fr/archives.htm
Project Gutemberg:
    http://www.gutenberg.org/wiki/Main_Page

Internet pages with digitized documents of individual scientists.

Galileo:	http://galileo.rice.edu/index.html
Hartlib:	http://www.shef.ac.uk/library/special/hartlib.html
Boyle:	http://www.bbk.ac.uk/boyle
Newton:	http://www.newtonproject.sussex.ac.uk/prism.php?id=1
Einstein:	http://www.alberteinstein.info/
Darwin:	http://darwin-online.org.uk
Leonardo da Vinci:	http://www.leonardodigitale.com/login.php
Lamarck:	http://www.lamarck.cnrs.fr/
Linnaeus:	http://linnaeus.c18.net/Doc/presentation.php
Lavoisier:	http://moro.imss.fi.it/lavoisier
Leibnitz:	http://leibnizviii.bbaw.de/
Buffon:	http://www.buffon.cnrs.fr/
Ampère:	http://www.ampere.cnrs.fr/
Galvani:	http://cis.alma.unibo.it/galvani/textus.html

Annex 2

Websites of some museums providing multimedia resources of history of science.

Conservatoire Nationale des Arts et Métiers (Paris) :
   www.arts-et-metiers.net/musee.php?P=191&lang=fra&flash=f
Museo Galileo (Firenze)
   http://www.museogalileo.it/en/index.html
The History of Science Museum at Oxford University
   http://www.mhs.ox.ac.uk/index.htm
Ingenious (National Museum of Science and Industry, London)
   http://ingenious.org.uk/
Pavia Project Physics (Università degli Studi di Pavia)
   http://ppp.unipv.it/PagesIT/pppit.htm
Making the Modern World (National Museum of Science and Industry, London)
   http://www.makingthemodernworld.org.uk/
Museo Nacional de Ciencia y Tecnología (Madrid)
   http://www.mec.es/mnct/movimientos/CD/acceso.html

Annex 3

Websites of online courses for the history of science

Universidad Nacional de Educación a Distancia (UNED)

> http://www.uned.es/pfp-evolucion-historica-principios-quimica/

Open University (OU)

> http://www3.open.ac.uk/courses/classifications/humanities__arts__languages_
> _history-history_of_science__technology_and_medicine.shtm

Open Learn (OU)

> http://www.open.ac.uk/openlearn/home.php

---

[1] Concerning this same topic, it is necessary to mention the opinion article by Sage Ross published in a *Newsletter of the History of Science Society* (Sage 2008: 37;1, 1/6) about the relevance of the digital encyclopaedia *Wikipedia* in relation to the history of science. In this regard, the historian of science Alan Rocke called upon members of the *History of Science Society* to review *Wikipedia* articles on historical material that contain flagrant errors (Rocke 2008, 37; 2, 3)

[2] In this respect, it is worth remembering the *Internet History Source Book Project*, http://www.fordham.edu/halsall (accessed 28 Feb. 2011) and the portal ECHO (*Exploring and Collecting History Online*) http://echo.gmu.edu/index.php (accessed 28 Feb. 2011) as a search engine designed for seeking information on science, technology and industry. Unlike the general search browsers (such as Google), which may provide hundreds of references for any item, this search engine can perform more selective searches.

[3] Conceptions about the nature of science are as varied as the approaches that have been made around this enterprise called "science". This means that we can often find complementary ideas on the nature of science, depending on the point of view of philosophy, sociology, the history of science or even cultural anthropology. No single conception of the nature of science exists, although interesting attempts have been made in science education to dismiss certain mythical misunderstandings of the nature of science (McComas 2000).

[4] The former Istituto e Museo di Storia della Scienza until June 11, 2010

[5] From the academic year 2009-2010, there is also a course for postgraduate students intended for training pre-service secondary education science and mathematics teachers. The History of Science is a compulsory part of this course and the time allocated to it is a matter left to the decision of each university.

[6] This course has been a collaborative effort by Agustí Camós (UAB), Pere Grapí – coordinator (UAB), Maria Rosa Massa (UPC), Maria Fàtima Romero (UPC) and Raimon Sucarrats (UAB).

[7] The following URLs have been accessed on 28 Feb. 2011

The Role of the History of Mathematics in Teacher Training Using ICT

Mª Rosa Massa-Esteve

Departament de Matemàtica Aplicada I. Universitat Politècnica de Catalunya; Spain;
m.rosa.massa@upc.edu

ABSTRACT: As an explicit and implicit resource in the classroom, the history of mathematics enables the learning of mathematics to be improved. The history of science, and specifically the history of mathematics, is a very fruitful tool to convey the perception of Mathematics as a useful, dynamic, humane, interdisciplinary and heuristic science. Furthermore, the analysis of significant original sources in the classroom constitutes a powerful tool for understanding mathematics. In this sense, the analysis of the contributions of history of mathematics to teacher training is highly relevant. In this chapter, we discuss the principal features of teacher training in the history of mathematics through the syllabus and development of an online course for in-service teachers of mathematics.

Introduction

The history of mathematics shows how mathematics has frequently been used to solve problems concerning human activity as well as for helping to understand the world that surrounds us. The study of historical processes enables us to see how the different aspects of mathematics have been combined together in a repeated interaction of application and development. Thus, for example, geometry, which emerged as a means of measure, has evolved alongside the problems of measurement (see Stilwell, J., 2010 (first ed. 1989)); trigonometry has developed in order to solve problems of both astronomy and navigation (see Zeller, 1944 and Maor, 1998), while algebra, which came more to the fore in problem-solving, especially in mercantile arithmetic during the Renaissance, was later to become an indispensible tool for solving problems in geometry and number theory (see Bashmakova & Smirnova, 2000 and Massa-Esteve, 2001 & 2005a), etc..

All this knowledge will undoubtedly enrich the mathematical background and training of teachers[1]. Furthermore, as we analyze in the following section, the history of mathematics can be employed both implicitly and explicitly. Learning about the history of mathematics can therefore contribute to improving the integral education and training of students.

Research on this subject is not new; indeed, in Catalonia[2] the implementation of the history of mathematics in the classroom has for twenty years inspired some individual initiatives among teachers. For instance, since the academic year 1990/91 research funding extended annually to teachers by the Catalan Government Department of Education has been devoted to research into the relations between the history of science (including mathematics) and its teaching. This research has resulted in the drawing up of reports that are now available to other teachers.

The creation of the Barcelona History of Mathematics Group (ABEAM) in 1998[3] also constituted a significant step forward. The aim of this group of teachers of Mathematics is to develop History of Mathematics materials to be used in the classroom[4].

As a collective initiative we may also mention that from 2003 to the present, Pere Grapí and the author of this chapter have been coordinating workshops on the History of Science and Teaching organized by the Catalan Society of History of Science and Technology (SCHCT), and are subsequently overseeing the publication of the proceedings[5]. The aim of these workshops is to enable teachers to share their experiences in the classroom as well as to discuss the criteria and conditions for these implementations. More than 50 people participate in these workshops, including historians of science and teachers who are interested or already practicing the implementation of history of science in science [math] classrooms. Many of these participants are from Spain, although sometimes people from France and Portugal also attend. The presentations, which number more than 20, deal with various aspects of the subject: the experiences and results of historical activities used by teachers in the classroom; analysis of the role of the history of science in teacher training and in the institutional curriculum, and more recently the use of ICT for introducing history of science in classroom. Each workshop also consists of a talk by a guest speaker to enable the audience learn from the experiences arising from the introduction of the history of science in classroom in other countries. For example, previous guest speakers have been: Peter Ellis (2003), Fabio Bevilacqua (2005), Peter Heering (2007), Ahmed Djebbar (2009), while this year, 2011, Michael Matthews will be the guest speaker.

In the academic year 2007-2008, the Catalan Government Department of Education introduced some compulsory elements of the history of science into the curriculum for secondary education. Specifically, the new mathematics curriculum for secondary schools in Catalonia, published in June 2007, contains notions of the historical genesis of relevant mathematical subjects within the syllabus[6].

Furthermore, the academic year 2009-2010 saw the inauguration of a new course for training pre-service teachers of mathematics in secondary education. The syllabus of this Master's degree launched at the universities includes a compulsory section on the history of mathematics and its use in the classroom[7].

In the same academic year 2009-2010, an online course on the history of science for in-service science teacher training was also set up under the name of "Science and Technology through History"[8]. During the last academic year 2010-2011, we have again conducted a renewed version of this online course for in-service mathematics teachers in the *Institute of Science of Education* (ICE) at the Universitat Politècnica de Catalunya. The course has been renamed: "The History of Mathematics and Science for Secondary Education", and furthermore a third edition will be launched in the current academic year 2011-2012. In this chapter we discuss the principal features of teacher training in the history of mathematics through the syllabus and development of the Mathematics block of this online course. First, in the following section we analyze our theoretical and practical approach on contributions of the history of mathematics in the classroom, which form the core of this course.

Usefulness of the history of mathematics in the classroom

We describe the usefulness of the history of mathematics in the classroom through our theoretical and practical approach, with the aim of persuading teachers about the need for this type of training. Knowledge of the history of mathematics can assist in the enrichment of teaching tasks in two ways: by providing students with a different vision of mathematics, and by improving the learning process.

A different vision of mathematics

Teachers with knowledge of the history of mathematics will have at their command the tools for conveying to students a perception of this discipline as a useful, dynamic, human, interdisciplinary and heuristic science[9].

- *A useful science.* Teachers should explain to students that mathematics has been an essential tool in the development of different civilizations. It has been used since antiquity for solving problems of counting, for understanding the movements of the stars and for establishing a calendar. There are many examples right down to the present day in which mathematics has proved to be vital in spheres as diverse as computer science, economics, biology, and in the building of models for explaining physical phenomena in the field of applied science, to mention just a few of the applications.

- *A dynamic science.* It is also necessary whenever appropriate to teach students about problems that remained open in a particular period, how they have evolved and the situation they are in now, as well as showing that research is still being carried out and that changes are constantly taking place.

- *A human science.* Teachers should reveal to students that behind the theorems and results there are remarkable people. It is not merely a question of recounting anecdotes but rather that students should learn something about the mathematical community; human beings whose work consisted in providing us with the theorems we use so frequently. Mathematics is a science that arises from human activity, and if students are able to see it in this way they will probably perceive it as something more accessible and closer to themselves.

- *An interdisciplinary science*. Wherever possible, teachers should show the historical connections of mathematics with other sciences (physics, biology, engineering, medicine, architecture, etc.) and other human activities (trade, politics, art, religion, etc.). It is also necessary to remember that a great number of important ideas in the development of science and mathematics itself have grown out of this interactive process.

- *A heuristic science*. Teachers should analyze with students the historical problems that have been solved by different methods, and thereby show them that the effort involved in solving problems has always been an exciting and enriching activity at a personal level. These methods can be used in teaching to encourage students to take an interest in research and to become budding researchers themselves[10].

Teachers with knowledge of the history of mathematics can show students a further relevant feature of mathematics – that it can be understood as a cultural activity. History shows that societies develop as a result of the scientific activity undertaken by successive generations, and that mathematics is a fundamental part of this process. Mathematics can be presented as an intellectual activity for solving problems in each period. The societal and cultural influences on the historical development of mathematics provide teachers with a view of mathematics as a subject dependant on time and space and thereby add an additional value to the discipline.

It is also worth pointing out that not only as teachers, but also as mathematicians, the history of mathematics enables us to arrive at a greater comprehension of the foundations and nature of this discipline. The history of mathematics provides the devotees of this science with a deeper approach to an understanding of the mathematical techniques and concepts used every day in the classroom. It helps to understand how and why the different branches of mathematics have taken shape: analysis, algebra, geometry, etc., their different interrelations and their relations with other sciences.

An improvement in the learning process

The history of mathematics as a didactic resource can provide tools to enable students to understand mathematical concepts better. The history of mathematics can be employed in the mathematics classroom as an implicit and explicit didactic resource[11].

The history of mathematics as an implicit resource can be employed by teachers in the design phase by choosing contexts, by preparing activities (problems and auxiliary sources) and also by drawing up the teaching syllabus for a concept or an idea[12].

In addition to its importance as an implicit tool for improving the learning of mathematics, the history of mathematics can also be used explicitly in the classroom for the teaching of the mathematics. Although by no means an exhaustive list, we may mention four areas where the history of mathematics can be employed explicitly in the Catalonia educational system.

First of all, history can be employed explicitly in the content of compulsory research work undertaken by students in Catalonia in their second year of Baccalaureate (18 year-olds). History situates students in a more general context, since problems are addressed within a global framework of mathematics and within the overall field of science.

The list of titles of research works dealing with the history of mathematics proposed by mathematics teachers is somewhat extensive, but as examples we may quote the following: Pythagoras and Music; The Golden Mean; On Fermat's Theorem; Pascal's Arithmetic Triangle as a Tool for Resolution; Perspective and its History in the Work of Leonardo da Vinci, Luca Pacioli and Albert Dürer; Women and Science; On Incommensurability: A Mathematical and Philosophical Problem, etc. The majority of these research projects have already been completed by Baccalaureate students, while some of them have been published (see Granados *et al.*, 1998) and received an award from the Research Department of the Catalan Government. They not only show the historical evolution of an idea or a concept, but also involve mathematical research enabling students to become familiar with mathematical reasoning from other periods and cultures, as well as in other contexts. The mathematical work that can be undertaken in each of these research fields is highly diverse and range from very simple problems to more complicated proofs. The mathematical learning entailed in such research works will depend as much on students' efforts and motivation as on the specific guidance provide by teachers.

Secondly, the history of mathematics can be employed explicitly in the creation of elective credits, which may arouse interest in mathematics in even the most unmotivated students. For instance, we may mention the variable credit aimed at 4th year secondary school students in Catalonia: "A Journey through the History of Squaring the Circle". This credit can furnish students with a good introduction to the history of mathematics through the learning of geometry according to theorems by Tales and Pythagoras, the golden mean, the geometry of the circle, the inscription and circumscription of polygons, the different approaches to the number π, and so on.

Thirdly, the holding of workshops, centenaries and conferences provides further types of activities in which history can be used explicitly to achieve a more comprehensive learning experience for students. For instance, the workshop devoted to the study of the life and work of René Descartes (1596-1650), held in 1996 at INS Carles Riba (a high school in Barcelona), afforded students additional education from a mathematical, philosophical, physic or historical perspective[13]. In 17th century Europe, subjected as it was to exhausting religious conflicts but with the intuition that a new era had begun, human knowledge was opening up to new exploratory possibilities. Philosophy, mathematics, physics, music and language were to become the facets of a new geometrical configuration making up modern rationality. In the course of this workshop, students discovered from an interdisciplinary perspective how Descartes' personality and lifetime achievements, institutions and historical events during his life shaped an intellectual atmosphere generating questions and reflexions arising from the same problems and concerns[14].

Finally, the explicit use of significant original sources in the classroom is the activity that can provide students with more valuable means for a better understanding of mathematical concepts[15].

Historical texts can be used throughout the different steps in the teaching and learning process: to introduce a mathematical concept; to explore it more deeply; to explain the differences between two contexts; to motivate study of a particular type of problems or to clarify a process of reasoning.

In order to use historical texts properly, teachers are required to present historical figures in context, both in terms of their own objectives and the concerns of their period. Situating authors chronologically enables us to enrich the training of students. Thus, students learn different aspects of the science and culture of the period in question in an interdisciplinary way. It is important not to fall into the trap of the amusing anecdote or the biographical detail without any mathematical content. It is also a good idea to have a map available in the classroom to situate the text both geographically and historically.

Teachers should clarify the relationship between the original source and the mathematical concept under study, so that the analysis of the significant proof should be integrated into the mathematical ideas one wishes to convey. The mathematical reasoning behind the proofs should be analyzed and contextualized within the mathematical syllabus by associating it with the mathematical ideas studied on the course so that students may see clearly that it forms an integral part of a body of knowledge. In addition, addressing the same result from different mathematical perspectives enriches students' knowledge and mathematical understanding.

In order to transmit to the students the idea that mathematics is a science in a continuous state of evolution, and that it is the result of the joint and ongoing work of many people rather than knowledge amassed by independent contributions arising from flashes of inspiration, it is necessary for teachers themselves to be well trained in the history of mathematics. Furthermore, all these implicit and explicit contributions of the history of mathematics in the classroom lead us to the conclusion that mathematics teachers could enrich their educational tasks if they are conversant with the genesis and evolution of mathematical ideas and concepts. It is a well-known fact that the history of mathematics scarcely ever forms part of the training of in-service teachers. Indeed, for many years there has been no compulsory subject related with the history of science in the training of pre-service mathematics teachers, and thus many in-service teachers have never had the opportunity to learn about the history of mathematics as part of their educational background. It is therefore our aim to fill this training gap for in-service mathematics teachers by developing and putting into practicing an online course on the history of mathematics.

An online course on the history of mathematics for in-service mathematics teachers

By means of the programming, development and putting into practicing of an online course on the history of mathematics for in-service teachers, we analyze the main features of the training requirements for teachers of mathematics.

The contents of the course (see Figure 1) can be found on a website designed for teacher-training and is managed through the Moodle learning management system; material for each session is available in extracts from original texts or secondary sources to enable teachers to analyze them according to specific guidelines[16].

The use of a website allows teachers to work freely without conforming to a fixed timetable. Teachers answer the questions and the tutor corrects and clarifies the exercises. The advantage of e-learning by internet is that it allows direct communication both with the tutor and other teachers.

```
D000: Ciència i tècnica a través de la història
Índex dels blocs i mòduls

• Bloc comú.
    ○ Mòdul 1. Desenvolupament i objectius de la història de la ciència i de la tècnica
    ○ Mòdul 2. Estructura i organització de la història de la ciència i de la tècnica
    ○ Mòdul 3. Les fonts de la història de la ciència
    ○ Mòdul 4. Història de la ciència i ensenyament
• Bloc específic de biologia i geologia.
    ○ Mòdul 1. Les classificacions de Linné i Buffon
    ○ Mòdul 2. La cèl·lula i la teoria cel·lular
    ○ Mòdul 3. L'evolució
    ○ Mòdul 4. Les grans controvèrsies geològiques
• Bloc específic de física i química.
    ○ Mòdul 1. La matèria
    ○ Mòdul 2. Els canvis a la matèria
    ○ Mòdul 3. L'Univers
    ○ Mòdul 4. Les interaccions a la matèria
• Bloc específic de matemàtiques.
    ○ Mòdul 1. Les matemàtiques a l'antiguitat
    ○ Mòdul 2. De la ciència àrab a les àlgebres renaixentistes
    ○ Mòdul 3. La Revolució Científica
    ○ Mòdul 4. La matemàtica moderna
```

Fig. 1: Main page of online course

The course (see Figure 1) is divided into one common block, three specific blocks (biology and geology, physics and chemistry, and mathematics), and a final project. The common block is compulsory for all the students and its general objective is to provide elements of the historiography of science to enable teachers to deal better with the history of science. Upon completion of this common block, teachers must choose one specific block according to their academic training or to their teaching practice. The course lasts for four months and is recognized by the educational authorities as a 45-hour teacher training course. There are also two face-to-face sessions, one at the beginning for the presentation of the course, and the other at the end for showing and discussing the final projects containing the experience of historical texts implemented in the classroom by teachers.

Goals and contents of mathematics block

The aim of this specific block is to explore the past of mathematics and thereby to trace the emergence and development over time of concepts, theorems, methods and axiomatic which we find in textbooks today. This is done according to a pragmatic, logical and didactic conception that often does not coincide with the historical order in which such concepts and theorems were either discovered or invented. This general aim can be broken down into four particular goals corresponding to the different facets of the course:

1. Learning about the sources on which knowledge of mathematics in the past is based. This involves reading and interpreting some selected classical mathematical texts, as well as learning how to locate and use historical literature.
2. Recognising the most significant changes in the discipline of Mathematics; those which have influenced its structure and classification, its methods, fundamental concepts and its relation to other sciences.
3. Showing the socio-cultural relations of mathematics with politics, religion, philosophy and culture in each period, as well as with other spheres.
4. Encouraging teachers to reflect on the development of mathematical thought and the transformations of natural philosophy.

As regards specific mathematical content, since the course is a general history of mathematics, the intention is not to address all subjects or to study all authors in depth. However, a careful selection of both subjects and authors has been made, which should be complemented by individual final projects to be handed in to teachers on conclusion of the course[17]. Indeed, not all historical texts are suitable for use in the mathematics classroom. The criteria for the initial selection include significant historical texts related to compulsory historical contexts in the new Catalan curriculum. Historical texts (e.g. proofs or problems) should in some way be anchored in the appropriate mathematical context. We therefore analyze relevant historical texts in order to enable teachers to understand fully the origin and evolution of different areas of mathematics such as algebra, trigonometry, probability theory and infinitesimal calculus. Four main periods in the history of mathematics are regarded as essential and will be addressed chronologically: Mathematics in Antiquity; From Arab science to Renaissance algebra; the Scientific Revolution, and Modern Mathematics. These periods will be presented as modules, and each one will be divided into three sections (in the course called three "activities") in which contributions made by different civilizations will be studied. The table below shows the different subjects of mathematics in the course.

Modules	Mathematics in Antiquity.	From Arab science to Renaissance algebra.	Scientific Revolution.	Modern Mathematics.
First activity	Cuneiform Tablets.	Contributions of Arab science. The Beginnings of Algebra.	The Development of Plane Trigonometry.	Probability Theory.
Second activity	Egyptian Papyrus.	The Beginnings of the Probability Theory. Games of Chance.	The Algebraicization of Mathematics.	The Fundamental Theorem of Algebra.
Third activity	Greek Geometry.	Mercantile Arithmetic. The Resolution of Equations in the Renaissance.	The Beginnings of Infinitesimal Calculus.	The Concept of Function. Exponential and Logarithmic Function.

For instance, in the first module, "Mathematics in Antiquity", the three activities are as follows: cuneiform tablets, Egyptian papyrus and Greek geometry. Figure 2 shows the main page of the first module of mathematics.

In each module, and under the title "to know more", one can find specific bibliographic references useful for improving the knowledge of the subject in each module.

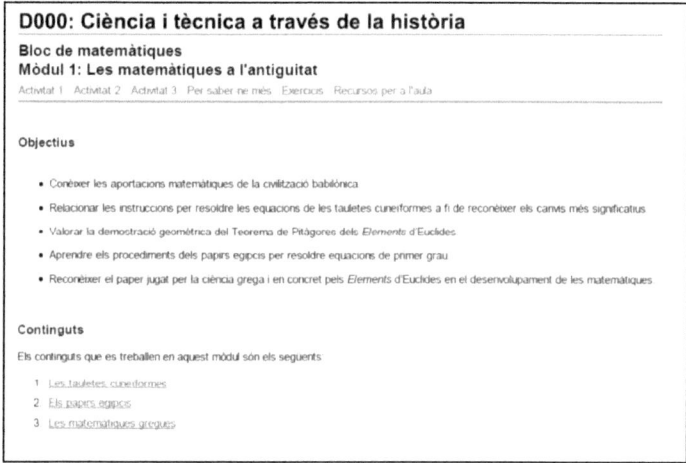

Fig. 2: Goals and contents of the first module of mathematics: "Mathematics in the Antiquity".

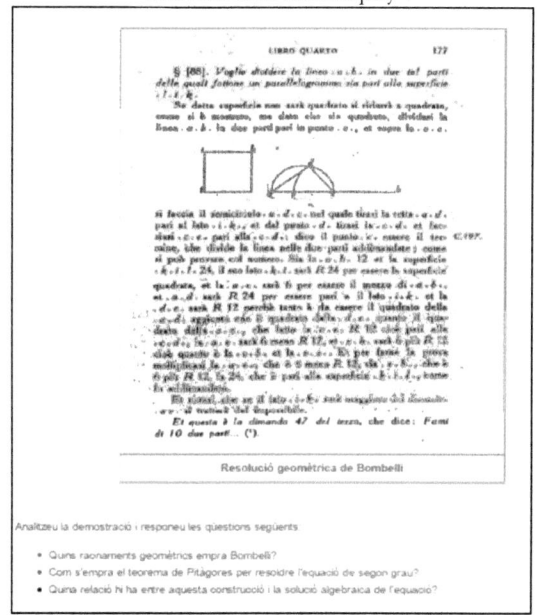

Fig. 3: Practical exercise on *Algebra* (1572) by Bombelli

The method employed for preparing the course material falls within the line of historical research whose purpose is to understand the genesis and evolution of mathematical concepts in their own context. This process should be understood in each period, in terms of mathematical knowledge and the intentions brought to bear, rather than in terms of what was to happen later. A further relevant question is the use of original sources (see Figure 3) as well as proofs suitably prepared to provide new knowledge and fresh ideas.

Practical exercises, internet forums and resources for mathematics classroom

The subjects are generally addressed in a twofold manner, with one part consisting of exposition and the other of practical exercises, the exposition sometimes being complemented by an internet forum.

For instance, in the different modules of the course we give an exposition of a brief historical journey through the development of algebraic equations[18]. While it is possible to deduce an algorithm for solving equations of second degree from Babylonian tablets (1800 BC) (see the first activity of the first module entitled "cuneiform tablets"), it was the Arabs who took the decisive step in the development of algebra. The mathematician, astronomer and member of the House of Wisdom in Baghdad, Mohamed Ben-Musa al-Khwarizmi (850 AD), is considered to be the creator of the rules of algebra. In his work *Hisâb al-jabr wal-muqqabala* (813 and 830), he classified equalities (now called equations) up to the second degree according to six different types, as well as explaining the method for solving them. It was Leonardo de Pisa, son of Bonacci (1180-1250), better known as Fibonacci, who disseminated all this knowledge in the West. Many of the problems dealt with in the algebra of the Arabs are to be found in his work *Liber abaci* (1202), as well as methods for calculating Hindu numeration (see the first activity of the second module entitled "Contributions of Arab Science. The Beginnings of Algebra"). The sketchiest period in the development of algebraic equations corresponds to the 13th and 14th centuries, when commercial mathematics flourished with the *Mercantile Arithmetics*, works that are still being explored and analyzed. The knowledge of these mercantile arithmetics and Arab algebra were collected in a work by Luca Pacioli (1447-1517) entitled *Summa de Arithmetica, Geometria, Proportioni & Proportionalità* (1494), which was widely known at that time. Later, Girolamo Cardano (1501-1576) and Rafael Bombelli (1526-1573), among other algebraists of the *Cinquecento*, also contributed with their respective works, *Artis Magnae sive de Regulis Algebraicis* (1545) and *Algebra* (1572), to the solution of cubic and quartic equations with syncopated algebra (see the third activity of the second module entitled "Mercantile Arithmetic. The Resolution of Equations in the Renaissance"). In this brief historical review, the work *In Artem Analyticen Isagoge* (1591) by François Viète (1540-1603) constitutes a landmark for its use of symbols, not only for representing unknown quantities but also known quantities. It included new algebraic procedures and emphasized their use for solving equations in arithmetic, geometry and trigonometry. Notwithstanding, the most influential figure was René Descartes (1596-1650) with his work *La Géométrie* (1637), in which the notation was that in current use today, except for two minor variations: the equals sign and the square. Descartes constructed an algebra of segments, and in the construction he makes of the solution to the second degree equation there appears the formula we still employ today (see the second activity of the third module entitled "The Algebraicization of Mathematics").

The course is mainly structured around the practical exercises (two for each activity), which are based on the exposition of texts to be analyzed, and around questions for guiding the reading and analysis of the texts.

For instance, in order to exemplify some of the materials, in the activity on cuneiform tablets there are two practical exercises, one for showing how the scribes of Babylon approximate the square root, and the other on the interpretation of the instructions for solving a problem in a Babylon tablet by setting up a second degree equation. Furthermore, in the third activity of the second module, the practical exercise on the geometrical construction by Bombelli (see Figure 3) requires teachers to analyze his proof and answer some questions: "What geometrical reasoning did Bombelli use? What is the role of Pythagoras' theorem in solving the equation of second degree? What relation is there between this construction and the algebraic solution of the second degree equation?." These questions enable teachers to consider the solution of second degree equations from a geometrical point of view, as well as prompting teachers to reflect on the relation between algebra and geometry through history.

Each module also includes four prescriptive exercises to be sent to the tutor. The solutions to these exercises-problems are derived from the reading of historical texts[19]; for instance, solving problems similar to those dealt with in the historical text, replicating significant historical mathematical proofs, commenting on a paragraph selected from an historical text, etc.

Some of these prescriptive exercises involve open internet forums in which the subject is debated according to a preset list of questions; any doubts that may arise are clarified in subsequent discussions with the tutor and with other teachers; for example, an internet forum discussion on the significance of Descartes' *Géométrie* in the activity on algebraicization of mathematics was particularly exciting. The educational nature of the course is conducive to discussion and communication in an atmosphere of collaborative e-learning.

Each module in the course also provides mathematical activities (under the title "resources for the classroom") for use in the classroom. Since the history of mathematics is currently a compulsory subject in the new curriculum in Catalonia, this course also includes various classroom activities that have already been tested in different schools with satisfactory results, according to reports from teachers. There is no systematic or organized evaluation of each mathematical resource presented in the course; we have only individual questionnaires in which students have consistently expressed a great interest in learning about the history of science (in particular, the history of mathematics). Wherever possible, resources have been designed in relation to mathematical texts from the different periods under study as a resource for use in the classroom. These resources are considered to be those which offer the greatest motivation for students, in addition to providing a different way of learning new mathematical concepts. The resource activities to be carried out have been drawn up in the form of a dossier to enable students to become familiar with the concepts by means of a constructive learning method. The aim is to inter-relate the different parts of mathematics, such as geometry, trigonometry and arithmetic, and to help students to recognise and employ the connections between mathematical ideas, thereby contributing to their mathematical education; for instance, the resource based on the demonstration of Pythagoras' theorem of Euclid's *Elements*, or on the demonstration of some propositions in Aristarchus of Samos' work (see Figure 4). In order to exemplify some of the materials, we present the latter, which deals with the work *On the Sizes and Distances of the Sun and Moon* (ca. 287 BC) by Aristarchus of Samos (ca. 310-230 BC). The guidelines for implementing this resource in the mathematics classroom advise teachers, to begin with a brief presentation of the epoch, Greek astronomy, and the person of Aristarchus himself, before giving students the dossier to complete. Teachers should then situate their work within the history of trigonometry and analyze the aims of the author as well as the features of the work. Finally, through the dossier students should be prompted to follow the reasoning of this demonstration, Proposition 7, in order to arrive at new mathematical ideas and perspectives. After the implementation of the dossier, teachers and students should analyze their findings together, in this case the mathematical strategies used by Aristarchus in this proof as well as the mathematical concepts under discussion. This resource is designed to be implemented in a mathematics classroom during the last phase of compulsory education (14-16 year-olds) as part of the introduction to trigonometry.

De fet demostra que 1/18 > sin 3° = CB : AB > 1/20. Mostrarem, tot seguit, la demostració de la primera desigualtat : 1/18 > CB : AB o sigui AB > 18CB.

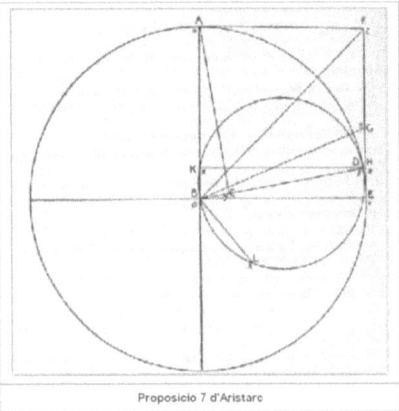

Proposició 7 d'Aristarc

Demostració

Sigui A el centre del Sol, B el centre de la Terra i C el centre de la Lluna, quan se'ns mostra partida per la meitat. Llavors CB representa la distància a la Lluna des de la Terra i AB representa la distància al Sol des de la Terra. Per a la hipòtesi núm. 4, l'angle BAC és 3°, llavors l'angle ABC que mesura l'allunyament de la Lluna al Sol és 87° ja que BCA és recte. Tot seguit, Aristarc dibuixa una circumferència de centre B i radi AB i estudia el problema en BHE, triangle semblant construït de costats perpendiculars al donat, o sigui que l'angle DBE val 3°. A més, completa el quadrat determinat pels costats AB i BE amb els costats AF i FE.

Sigui, doncs, l'angle FBE igual a 45° i l'angle GBE la meitat o sigui 90/4. Fent la raó entre els dos angles GBE i DBE, que val 3, dóna 15 és a 2.

Diu Aristarc que, com que sabem que la raó entre els costats oposats a aquests angles és més gran que la raó entre ells, podem escriure que

$$GE : HE > (GBE) : (DBE) = 15 : 2.$$

Ara, aplicant el Teorema de Pitàgores al triangle isòsceles (BE = FE) format per la meitat del quadrat, es compleix que $FB^2 = 2 BE^2$. A continuació aplica proporcions als triangles semblants, arribant a la conclusió que $FG^2 = 2 GE^2$.

L'estratègia que fa servir tot seguit és utilitzar la raó 50 : 25 = 2 : 49 : 25. Llavors escriu

$FG^2 : GE^2 = 2 > 49 : 25$. Traient l'arrel quadrada queda FG : GE > 7 : 5. Component la raó (componendo), FG + GE = F E (Euclides, 1956 114-115) resulta

$$FE : GE > 12 : 5 = 36 : 15$$

Però com que abans havia demostrat que GE : HE > 15 : 2, fent el producte de les dues raons (ex aequali), FE : GE amb GE : HE, resulta

$$FE : HE > 36 : 2 = 18 : 1$$

O sigui que FE > 18 HE, però com que FE = BE (costats del quadrat) llavors BE > 18 HE. Sabem també que BH que és la hipotenusa és més gran que BE que és un catet, llavors

Fig. 4: Aristarchus' proof of Proposition 7, which deals with the ratios between the distance of the Sun and the Moon from the Earth.

Final project

It is worth mentioning that a key part of the course is the final project that teachers should deliver in writing, as well as an oral exposition in the final session, which is conducted face-to-face. The presentation of all this work and the discussion between teachers provide a rich source of knowledge. This final project should be based on an author or a text chosen by teachers, and enables students to follow particular procedures and learn mathematical concepts from a different perspective.

Teachers are required to prepare a resource including a dossier for the students, which should be implemented in the classroom, as well as a dossier for the teachers themselves. In these dossiers for teachers there is a file containing the level of student achievement, the aims of the resource, the knowledge of the contents required for carrying out the activity, the indicative timing, the skills to be acquired by students, the evaluation criteria and the process of implementation in the classroom.

All final works implemented in classroom are subject to a twofold evaluation: one to determine students' knowledge, and other to assess the activity. This second evaluation is used to improve the content of future activities. The form is shown in Figure 5 and evaluates the following: the understanding and usefulness of the activity; the structure of the contents, the clarity of the explanations and the degree of difficulty. The questions on this form read as follows: 1) Which aspect of activity did you like the best? 2) Which did you like least? 3) Would you leave anything out of the activity? 4) Would you add something to the activity? 5) Other remarks.

The subjects addressed in these final projects cover a broad range, some of which are as follows: The Arithmetic of Ancient Egypt, Thales' theorem, The Origins of Symbolic Algebra, Pythagoras' theorem, etc..

Avaluació de l'activitat

Indica el temps que has dedicat:

Valora del 0 al 10 els següents aspectes (0 = totalment en contra, 10 = totalment a favor) i respon a les següents preguntes:

Amb aquesta activitat he entès els conceptes que s'hi plantejaven.	
Aquesta activitat té una proporció de teoria i pràctica correcta.	
Crec que aquesta activitat té una dificultat elevada.	
L'estructuració dels continguts és correcta.	
Crec que la història de la matemàtica és útil per comprendre nous conceptes.	

1. Què és el que més t'ha agradat?

2. Què és el que menys t'ha agradat?

3. Trauries alguna cosa de l'activitat?

4. Afegiries alguna cosa a l'activitat?

5. Altres observacions:

Fig. 5: Model of evaluation of activity.

As an example of the materials used, we describe the design, implementation and assessment of a specific final project by a mathematics teacher: "The Arithmetic of Ancient Egypt", which deals with some arithmetic problems solved in the Rhind Papyrus (1.650 BC)[20]. The Rhind Papyrus included 87 mathematic problems on Arithmetic, Geometry and Algebra. This classroom activity was implemented in the last phase of compulsory education (14-16 year olds). Its design includes a brief presentation for other teachers describing the aims and mathematical skills acquired by students who participated in this activity.

The process of implementation in the mathematics classroom is as follows: the first hour is devoted to a power point presentation given by the teacher on Egyptian culture and the Rhind Papyrus; in the second and third hours, students are required to solve some arithmetic problems on Egyptian numeration, addition and subtraction as well as on Egyptian fractions; at the end of third hour students must complete a questionnaire for evaluation of the activity similar to the one below.

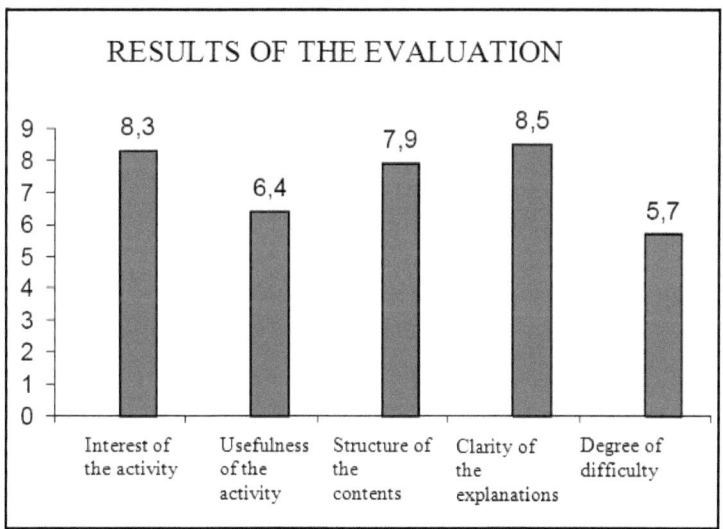

Fig. 6: Evaluation of the final project "The Arithmetic of the Ancient Egypt".

Assessment of this activity therefore includes a questionnaire on the level of interest shown in the activity (8,3), its usefulness (6,4), the structure of the contents (7,9), the clarity of the explanations (8,5) and the degree of difficulty (5,7) (see Figure 6). Most students requested more activities on mathematics from other cultures such as China, or on Egyptian geometry.

The teacher who implemented the activity reported that she was very satisfied with the results, adding that students applied themselves to the task with enthusiasm and rigor.

Concluding remarks

In order to ensure that the implementation of the history of mathematics in the classroom is a fruitful experience, the history of mathematics itself should form part of teacher training.

We therefore embarked on the preparation of an online course on the history of mathematics for in-service teachers. Use of the internet allows teachers to work in their classrooms and to participate in this online course without having to be present at the University. They will be able to work and e-learn freely from their computer at home, while at the same time connecting both with the tutor and with other teachers.

On conclusion of the course, final reports from some teachers reveal that although on occasion they had already addressed historical topics in their classroom, the course provided them with original sources and new ideas for future use in the mathematics classroom.

The importance of these kinds of courses for in-service teachers is unquestionable, since they provides a new and more comprehensive perspective on the subject, thereby enabling teachers to offer a much broader scientific training to their students. The analysis of significant original sources is vital for the overall education of students, providing them with additional knowledge of the social and scientific context of the different periods in which the texts were written. In addition, the use of cases taken from history is one of the resources that can be employed to improve the transmission of mathematical knowledge and achievement. It also provides a stimulus and incentive for those students who may lack mathematical motivation. Throughout this course, teachers acquire a more inclusive vision of the development of mathematics, as well as fresh didactical resources, both implicit and explicit, to enable them to improve their teaching of mathematics.

Thanks to courses of this type, teachers are able to extend their own knowledge, students are able to participate in a more enriching learning process, and the quality of teaching in mathematics as a whole is also improved.

References

Aristarco (2007), *Aristarco de Samos. Sobre los tamaños y las distancias del Sol y la Luna*, Intr., trad. y notas de Mª Rosa Massa Esteve. Cádiz: Servicio de Publicaciones de la Universidad de Cádiz.

Bashmakova, I; Smirnova, G. (2000), *The Beginnings & Evolution of Algebra*, A. Shenitzer (Trans.), Washington: The Mathematical Association of America.

Barbin, E. (2000), Integrating history: research perspectives. In J. Fauvel & J. van Maanen, (Eds.): *History in mathematics education: the ICMI study*, Dordrecht: Kluwert, 63-66.

Calinger, R. (ed.) (1996), *Vita Mathematica. Historical research and Integration with teaching*, Washington: The Mathematical Association of America.

Delshams, A; Massa Esteve, M. R. (2008), "Consideracions al voltant de la Funció Beta a l'obra de Leonhard Euler (1707-1783)", *Quaderns d'Història de l'Enginyeria* IX, 59-82.

Dematté, A. (2006), Fare matematica con i documenti storici. Una raccolta per la scuola scondaria di primo e secondo grado, Trento: Editore Provincia Autonoma di Trento – IPRASE del Trentino.

Fauvel, J.; Gray, J. (eds.) (1987), *The History of Mathematics: A Reader*, Londres: MacMillan.

Fauvel, J.; Maanen, J. V. (eds.) (2000), *History in mathematics education: the ICMI study*, Dordrecht: Kluwer.

Granados, J.; Comas, J.; Massa, M. R. (1998), *Recull de Treballs de Recerca, Premi Cirit 1997*, Barcelona: Institut de Batxillerat Carles Riba.

Grapi Vilumara, P.; Massa Eesteve, M. R. (eds.) (2005), *Actes de la I Jornada sobre la Història de la Ciència i l'Ensenyament Antoni Quintana Marí*, Barcelona: Societat Catalana d'Història de la Ciència I de la Tècnica.

Grapi Vilumara, P.; Massa Eesteve, M. R. (eds.) (2007), *Actes de la II Jornada sobre la Història de la Ciència i l'Ensenyament Antoni Quintana Marí*, Barcelona: Societat Catalana d'Història de la Ciència I de la Tècnica.

Grapi Vilumara, P.; Massa Eesteve, M. R. (eds.) (2011), *Actes de la VI Jornada sobre la Història de la Ciència i l'Ensenyament Antoni Quintana Marí*, Barcelona: Societat Catalana d'Història de la Ciència I de la Tècnica.

Guevara, I.; Massa, Mª R. (2005), "Mètodes algebraics a l'obra de Regiomontanus (1436-1476)", *Biaix* 24, 27-34.

Guevara, I.; Massa, Mª R.; Romero, F. (2008a), "Enseñar Matemáticas a través de su historia: algunos conceptos trigonométricos", *Epsilon* 23(1y 2), 97-107.

Guevara, I.; Massa, Mª R.; Romero, F. (2008b), "Geometria i trigonometria en el teorema de Menelau (100 dC)", *Actes d'Història de la Ciència i de la Tècnica* 1(2), 39-50.

Jahnke, H. N. et al. (1996), *History of Mathematics and Education: Ideas and Experiences*, Göttingen: Vandenhoeck und Ruprecht.

Maor, E. (1998), *Trigonometric Delights*, Princeton: Princeton University Press.

Massa, M. R.; Comas, J.; Granados, J. (eds.) (1996), *Ciència, filosofia i societat en René Descartes*, Barcelona: Institut de Batxillerat Carles Riba.

Massa Esteve, M. R. (2001), "Las relaciones entre el álgebra y la geometría en el siglo XVII", *Llull* 24, 705-725.

Massa Esteve, M. R. (2003), "Aportacions de la història de la matemàtica a l'ensenyament de la matemàtica", *Biaix* 21, 4–9.

Massa Esteve, M. R. (2005ª), "Les equacions de segon grau al llarg de la història", *Biaix* 24, 4–15.

Massa Esteve, M. R. (2005b), L'ensenyament de la trigonometria. Aristarc de Samos (310–230 aC.). In P. Grapi & M. R. Massa (eds.): *Actes de la I Jornada sobre la història de la ciència i l'ensenyament*, Barcelona: Societat Catalana d'Història de la Ciència i de la Tècnica, 95–101.

Massa Esteve, M. R. *et al.* (2006), Teaching Mathematics through history: some trigonometric concepts. In M. Kokowski (ed.): *Proceedings of the 2nd International Conference of the European society for the History of Science*, Cracow: European Society for the History of Science, 150-157.

Massa Esteve, M. R. (2007), "Leonhard Euler (1707–1783): l'home, el creador i el mestre", *Mètode* 55, 35–38.

Massa Esteve, M. R. (2008), "Congrés Internacional 300 Aniversari Leonhard Euler (1707–2007): Barcelona, 20-21 de setembre de 2007", *Quaderns d'Història de l'Enginyeria* IX, 307-310.

Massa Esteve, Mª R. (2010), Understanding Mathematics through its History. In H. Hunger (ed.): *Proceedings of the 3rd International Conference of the European society for the History of Science*, Vienna: European Society for the History of Science, 809-818.

Massa Esteve, M. R. *et al.* (2011), Understanding Mathematics using original sources. Criteria and Conditions. In E. Barbin, M. Kronfellner, C. Tzanakis (eds): *History and Epistemology in Mathematics Education. Proceedings of the Sixth European Summer University*, Vienna: Verlag Holzhausem GmbH, 415-428.

Roca-Rosell, A. (2011), 10. Integration of Science education and History of Science: The Catalan Experience. In P. V. Kokkotas, K. S. Malamitsa and A. A. Rizaki (ed.): *Adapting Historical Knowledge Production to the Classroom*, Rotterdam: Sense Publishers, 141-150.

Romero, F.; Massa, M. R. (2003), "El teorema de Ptolemeu", *Biaix* 21, 31–36.

Romero, F.; Massa, M. R.; Casals, M. A. (2006), La trigonometria en el món àrab. "Tractat sobre el quadrilàter complet" de Nasir Al-din Altusi (1201-1274). In J. Batlló [et al.] (eds.): *Actes de la VIII Trobada d'Història de la Ciència i de la Tècnica*, Barcelona: Societat Catalana d'Història de la Ciència i de la Tècnica, 569–575.

Romero, F.; Guevara, I.; Massa, M. R. (2007), Els Elements d'Euclides. Idees trigonomètriques a l'aula. In P. Grapí & Mª R. Massa, (eds.): *Actes de la II Jornada sobre Història de la Ciència i Ensenyament "Antoni Quintana Marí"*, Barcelona: Societat Catalana d'Història de la Ciència i de la Tècnica, 113–119.

Romero, F. *et al.* (2009), "La trigonometria en els inicis de la matemàtica xinesa. Algunes idees per treballar a l'aula", *Actes d'Història de la ciència i de la Tècnica* 2(1), 427-436.

Serres, M. (1991), *Historia de las Ciencias*, Madrid: Cátedra.

Smith, D. E. (1959), *A Source Book in Mathematics*, New York: Courier Dover Publications.

Stillwell, J. (2010) (1ra ed. 1989), *Mathematics and Its History*, Berlin: Springer.

Struik, D. J. (1969), *A Source Book in Mathematics, 1200–1800*, Minnesota: Harvard University Press.

Zeller, S. M. C. (1944), *The Development of trigonometry from Regiomontanus to Pitiscus*, Michigan: University of Michigan, Ann Astor.

On-line course:

http://www.xtec.cat/formaciotic/dvdformacio/materials/tdcdec/index.html, last access November 1th, 2011.

[1] Historians of mathematics have pursued various research lines at an international level that investigate how to use knowledge of the history of mathematics effectively in the classroom in order to improve the teaching and learning of mathematics. For more, see, Barbin, E., 2000, Jahnke, H. N., 1996, Calinger, R., 1996, Fauvel &Maanen, 2000, Demattè, 2006 and Massa Esteve, 2010.

[2] In Spain, each Autonomous Community is in charge of its own secondary and graduate education, so we focus only on the implementation of history in the mathematics classroom in Catalonia. For a survey of the implementation of history in the science classroom in Catalonia, see also Roca-Rosell, 2011.

[3] The coordinator of the group is Mª Rosa Massa Esteve, and the other members of group are: Mª Àngels Casals Puit (INS Joan Corominas), Iolanda Guevara Casanova (INS Badalona VII), Paco Moreno Rigall (INS XXV Olimpíada), Carles Puig Pla (UPC) and Fàtima Romero Vallhonesta (Inspecció d'Educació). The group is subsidized by the *Institute of Science of Education* (ICE) of the University of Barcelona.

[4] The list of the texts implemented includes extracts from the following works: *On the sizes and distances of the Sun and Moon* by Aristarchus of Samos (ca. 310-230 BC) (Massa Esteve, 2005b; Aristarco, 2007); Euclid's *Elements* (300 BC) (Romero, Guevara, Massa, 2007); Menelaus' *Spheriques* (ca. 100) (Guevara, Massa, Romero, 2008a-2008b); *Almagest* by Ptolemy (85-165) (Romero, Massa, 2003); *The nine Chapters on the Mathematical Art* (s. I. AC) (Romero et al., 2009); *Traité du quadrilatère* by Nassir-al-Tusi (1201-1274) (Romero, Massa, Casals, 2006) and *Triangulis Omnimodis* by Regiomontanus (1436-1476) (Guevara, Massa, 2005).

[5] See Grapi &Massa, 2005, 2007 &2011.

[6] The list of these historical contexts includes: The origins of the numeration system; the introduction of zero and the systems of positional numeration; geometry in ancient civilizations (Egypt, Babylonia); initial approaches to the number π (Egypt, China and Greece); Pythagoras theorem in Euclid's *Elements* and in China; the origins of symbolic algebra (Arab world, Renaissance); the relationship between geometry and algebra and the introduction of Cartesian coordinates; the geometric resolution of equations (Greece, India, Arab World); the use of geometry to measure the distance Earth - Sun and Earth - Moon (Greece).

[7] For example, in the Universitat Politècnica de Catalunya the title is: "Elements of the history of mathematics for the classroom".

[8] This course has been a collaborative effort by Agustí Camós (UAB), Pere Grapí, Co-ordinator (UAB), Mª Rosa Massa (UPC), Mª Fàtima Romero (UPC) and Raimon Sucarrats (UAB).

[9] On this vision of mathematics see Massa Esteve, 2003.

[10] On the evolution historical of the methods see Massa Esteve, 2005a.

[11] Specific examples of the use of implicit and explicit resources for improving the teaching of mathematics can be found in Massa Esteve, 2003.

[12] The implicit use of the history of mathematics in the classroom is based on what is known as the "genetic principle". This principle states that individual development from an intellectual point of view, and therefore the individual process of acquisition of knowledge, is a reflection of the different historical stages in the evolution of human knowledge. As regards mathematics, the potentiality of this idea and the need to apply it

seriously to teaching has been endorsed by outstanding figures such as H. Poincaré and F. Klein, and in Spain by J. Rey Pastor, P. Puig Adam and Miguel de Guzman. It is worth mentioning that this principle is currently the object of much study, especially as regards its application to teaching. See Massa Esteve, 2010.

[13] See Massa, Comas and Granados, 1996.

[14] As in 1996, on the 400th anniversary of the birth of Descartes, in 2007 the 300th anniversary of the birth of Leonhard Euler (1707-1783) was celebrated, the mathematician and engineer to whom we owe most of the contributions to mathematical analysis. Euler was a great creator; through his contributions he opened up new fields and gave the world new symbols, such as the number "e" for representing the base of logarithms, the number "i" standing for complex numbers, the letter F and the brackets for functions; he also gave new definitions, new formulae, new polynomials (known as Euler polynomials), Euler integrals, Euler lines, etc. See Massa Esteve, 2007, 2008 and Delshams-Massa, 2008.

[15] See Massa Esteve, M, R. *et al.*, 2006 and idem., forthcoming.

[16] We use translations of relevant historical texts or, wherever possible, original sources. Some of these texts can be found, for instance, in Smith, 1959 and Struik, 1969.

[17] Fatima Romero has collaborated in the design of this content.

[18] See Bashmakova & Smirnova, 2000 and Massa, 2005a.

[19] Readings are taken from general history books such as those by Serres, 1991, Fauvel & Gray, 1987, among others.

[20] In the Rhind Papyrus, the scribe explained that there was a copy of another papyrus written 200 years previously which contained the mathematical knowledge acquired up to that time.

Conceiving classroom IBST ICT EHST resources: practice analysis in mathematics

Thomas de Vittori

Laboratoire de Mathématiques de Lens, Université d'Artois, France; thomas.devittori@euler.univ-artois.fr

ABSTRACT: The mix of IBST, EHST and ICT in a teaching situation raises new didactic questions that can be analyzed so that its pertinence and wealth appear more clearly. At the beginning of this chapter, the analysis of such a session brings to light a potentially reusable structure that can then be tried on another topic. This first part of the study points out the usefulness of a new theoretical frameworks elaboration discussed more fully in the latter part of this text.

Introduction

As expressed in the International Commission on Mathematics Instruction (ICMI) world report on the issues of the use of history in mathematics education: "having history of mathematics as a resource for the teacher is beneficial" (Fauvel – van Maanen (Eds.) 2000, p.1). In a more general way, according to most educators all over the world, the usefulness of history in science teaching is now assured. In many countries, official instructions invite teachers in their courses, to propose activities with or about the history of science. This new deal in education has led to profound changes in the French teacher training programs. Often built in collaboration with historians of science, the new university curricula now comprise a non negligible part of epistemology and the history of scientific concepts. Some of the main aims of such courses are to help students to enhance their up and coming professional skills by thinking about the definition of science, the acquiring of knowledge on different topics and periods, and an introduction to historical methodology. In mathematics, as in the other scientific fields, the inquiry based learning can become a very well-balanced convergence point between history, education and new technologies. What is explored in this text are some didactic research questions on the use of history of mathematics in an inquiry based teaching approach using ICT.

Context and first questions

Consequently to the institutional changes, the use of the history of science is increasingly present in classrooms. Nonetheless, most teachers do not feel at ease when proposing such activities. The introduction of history into science courses is a new request and for which teachers have not yet been prepared. Many ways to help teachers are possible and the community is very active. For decades, lots of articles about successful experiments in classrooms or about potentially good activities have been published. Thus, numerous resources are now available but it is often up to the teacher to find what was and will be pertinent. There is no doubt that passionate teachers will be able to gain the benefit of such documents, but what about the others? This simple question in fact opens a new field of research in teacher training. Didactic studies on the use of the history of science in education are now engaged in many European countries and, in the mathematics education field, a major part of the community agrees with their usefulness (Jankvist 2009, Arcavi – Tzanakis 2000, Siu – Tzanakis 2004). The history of mathematics can interact within the lesson in multiple ways that seem only to be limited by teachers' creativity. Nevertheless, while looking at the resources proposed, some similarities are appearing. A first study (de Vittori – Loeuille 2009) has shown that the ways the pupils enter the task can be characterized and can be grouped together into five (may be non exhaustive) main types. Two of those, the technical and the linguistic ways are presented above. As the other three will not appear in the following discussion, (the epistemological, the practical and the recreational ones) their identification in these preliminary works will not be detailed. At a research level, the didactic analysis of university courses of the history of science (teacher training sessions for instance) can give very useful tools for the elaboration of new activities for the classroom. One of the main difficulties in a teacher training session is to make historical goals and mathematics goals interact cleverly. A first possible link between both fields appears when students are invited to solve a mathematical problem which leads them to a historical question. For example, in a teacher training experiment, the students can be invited to work on *Il Saggiatore* passage on Earth-comets distance, and redo one of Galileo's demonstrations about the position of comets (sub-lunar or supra-lunar objects). This geometrical activity gives them the opportunity to think about the reason for such a proof, that is to say, to think about the historical context of a cosmological problem: the nature of the comets. If the aim of the training session is the learning of the historical method of contextualization, the mathematical task becomes a way of reaching that teaching goal. This way is the technical one. Complementary to this first case, a second type of entry is due to the massive use of texts as a resource for history of science training sessions. Old written documents are intrinsically a source of questions about language. Even if the text is chosen carefully in order to be easy to read, some words, or expressions have to be enlightened by their historical context. In such situation, students are often invited to read texts and to ask questions about what they do not understand. This work on the vocabulary helps them to have a better understanding of the context of a scientific concept. This is the linguistic way. Both entries, the technical and the linguistic one, have been identified in a university framework but they can be used as theoretical elements to analyze a classroom activity. This classification helps to take into account the structure of a mathematics lessons comprising historical contents and aims. For instance, in a previously studied situation in secondary school (de Vittori – Loeuille, 2009), the technical entry was used when pupils had to redo an old geometrical construction, and the linguistic task appeared when they spontaneously questioned the teacher about the word "square" written in an old language (*quarré* instead of *carré*). In the first step, the mathematical task leads pupils to think about the historical reasons for such a construction (Hellenistic geometry, rules and compass constructions...). In the second, an etymological question helps them to memorize the name, and the characteristics, of polygons with which the lesson dealt.

Exploratory study

With these theoretical elements, let us go back to the specific case of IBST, ICT and HST teaching situations. As above mentioned, many classroom activities have been described in education revues but not many combine the three aspects. One of them is nonetheless very well described in the French revue *REPERES – IREM* (n°57 – October 2004). The first step of the sequence consists in a problem solving task inspired by Al-Khwarismi's five squares' problem. Without any information on the historical aspect of this exercise, the pupils themselves have to find a way to get into the problem.

> "Five brothers and sisters have inherited five squares of land whose sides measure five consecutive integers. The lots are assembled into two groups: the three smaller lots are on one side of a path, and the two larger lands are on the other side. The surfaces of both sides of the path are equal. How is the dimension of each field to be found?"

Many pupils try random integers and in a collective discussion the teacher suggests the use of a spreadsheet. This second step creates an interaction between ICT and the IBST part of the activity. This way of integrating ICT is very usual and shows that it is not necessary to be totally innovative to create interesting sessions. With the help of the software the pupils are able to identify the equation $x^2 + (x+1)^2 + (x+2)^2 = (x+3)^2 + (x+4)^2$ that appears in this problem. This equation is equivalent to the second degree equation $x^2 = 8x + 20$. The pupils do not know the general method of how to solve such a trinomial and the teacher proposes to read them one of Al-

Khwarizmi's texts on this topic. In his text, the Arabic scholar explains how to solve such a type of equation and the pupils are invited to follow the solving algorithm. Of course, it works, and with the help of EHST the pupils find the solution. What is the structure of this sequence? The activity comprises two main parts. In the first one, IBST and ICT are blended into a technical entry that leads the pupils to engage themselves in the reading of the historical text. Like any old text, the one of Al-Khwarizmi is strangely written for a 21th century reader and this experience gives the pupils an opportunity to discover the history of mathematics. In the article, the teacher gives much information about the documentation he has been using in order to prepare the session and to give it its historical color. Thus, the sequence takes place in a historical environment given by the teacher. As it is very specific to such a type of session, this environment change is an important element of the historical approach that should be kept in mind for further analysis.

The sequence described was successful and the author of the article gives lots of details on the work done by the pupils. The activity is done around documents of different types and follows a specific structure that can be studied. In order to evaluate the weight of these choices (documents and structure), a new classroom test was organized in 2009. With the help of a secondary teacher skilled in such historical sessions, the same problem was proposed. Due to some constrains, the session planned was shorter than the original one. Nonetheless, the structure was the same, with first an IBST-ICT technical entry followed by a linguistic task on historical texts. Even if other documents were used, the aims of the session were respected. In this lesson, the teacher tries to engage pupils in a inquiry based on the five squares' problem. Contrary to the original activity, in the second test pupils had had only to use their calculator to explore the possible solutions. The pupils were very interested, involved, and according to the teacher, the sequence had been a complete success.

The second step in the experiment was to try to extrapolate this two steps' structure on another topic. The possible themes are numerous, and due to pedagogical coherence problems, plane sections were chosen. The mathematical aim of the session was to prove that the section of a cylinder cut by a plane perpendicular to its axis is a circle. For the first step, the pupils had to explore a 3D figure in which a plane could be moved in order to make the section appear (the source file of the Geospace figure is given in the appendix).

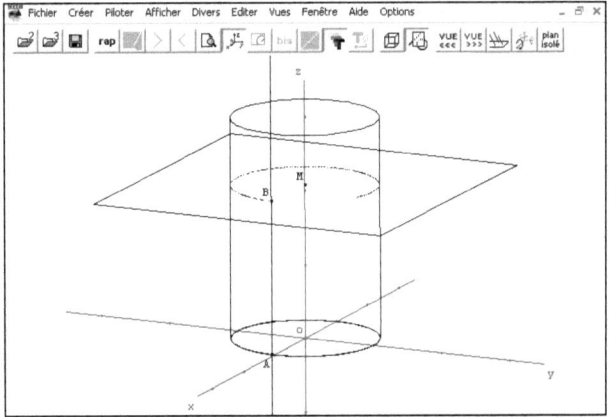

Fig. 1: Point M is movable (height of the plane). Point A is movable on the base circle and point B leaves a trace.

The software allows the rotating of the figure in the three directions. This function enables the pupils to check the nature of the section (circle or ellipsis). After the technical entry in this very classical problem (plane sections is one of the main research topics from Antiquity to Middle-Age), a historical text is given. The text is a quotation of Ibn al-Samh's fragment on the plane section (Rashed 1996, p.934). In his text, the Arabic scholar gives the main definitions and the proof of the most usual results on plane sections. In chapter 11, Ibn al-Samh proposes a demonstration on the plane section of a cylinder.

<11> Première espèce des sections de cylindre droit dont les bases sont deux cercles.

En coupant le cylindre par un plan parallèle aux bases on a une section, laquelle est donc engendrée par le mouvement d'une droite, dont une extrémité, fixe, est dans le plan de section, et qui pivote dans le plan jusqu'à ce qu'elle revienne à sa position initiale. La portion de plan balayée par cette droite s'appelle un cercle; ce que décrit l'autre extrémité s'appelle la circonférence. La droite mobile s'appelle le demi-diamètre. Le point fixe s'appelle le centre du cercle: toutes les droites qui en sont issues et qui vont jusqu'à la circonférence sont égales les unes aux autres. Cette section est nécessairement un cercle; en effet, nous relevons parmi ses propriétés qu'elle a un point intérieur, tel que toutes les lignes droites qui en sont issues et qui vont jusqu'à la périphérie de la section sont égales les unes aux autres; or, dans cette figure engendrée par le mouvement de la droite qu'est le cercle, nous trouvons un point, tel que toutes les droites qui en sont issues et qui vont jusqu'à la périphérie sont égales entre elles. Et si nous appliquons cette section sur le cercle dont le demi-diamètre est égal au demi-diamètre de ladite section, elle coïncide avec lui.

The pupils have to read the text and to use it in order to write their own demonstration. The text gives the opportunity to start a discussion and work on the definition of the geometrical objects (circle, straight line, segment, cylinder...). Moreover, both parts of the session point out the use of movement. The movement in geometry is an element of mathematics learning but it is also an interesting historical topic in itself. The session is followed by application exercises in which the movement is used to define geometrical figures (see appendix).

The elaborating process of the sequence on plane sections and geometrical movement follows the structure noticed in the first examples. IBST-ICT are mixed in a first technical entry that leads to an EHST work through a linguistic entry in an original historical source. Of course, the main aim of the lesson is on mathematics education. The teacher mainly teaches mathematics and history of science stays a secondary goal. Nonetheless, historical contents are transmitted and take up a great place in the elaboration phase. An important point is that in all the experiments, pupils had not really been directly in contact with historical aims. In fact, epistemological and historical knowledge goals are rather invisible for the students and below to the teaching situation. As explained in the activity about the five squares' problem, history is only rendered by the teacher's discourse ("Je présente rapidement ce mathématicien" says the author, Mercier (2004) p.56). Thus, pupils encounter the history of mathematics by the means of a specific environment driven by the teacher. As they are real teaching problems, the way the lesson has been created and the ways the teacher navigates from historical to mathematical environments can be regarded as didactic research objects which require specific studies.

Conclusion: Teaching with history of mathematics: elaboration tools and evaluation

In France, the history of mathematics in the classroom has been enhanced by a double institutional request. According to the government, firstly, teachers have to acquire knowledge on the history of science, and secondly, pupils have to think about the birth of concepts and their evolution. Of course, these recommendations have to take place in traditional teaching in which the learning of mathematical contents is the main purpose. Addressing what can appear as a paradox (how can a lesson with two different fields and learning aims be done?), in the aforementioned experiments, the answer was to deeply change the environment in order to make it historical. As previously shown above, several points can be modified to create such a situation. With a common structure, on two different topics, new modalities to enter the task have been used to reach simultaneously mathematical and historical goals. For instance, in the lesson on plane sections, in the IBST-ICT step, the technical entry helps pupils both to explore a geometrical situation and see by themselves the use of movement in geometry. Then the linguistic work made on the text leads to a think tank on the evolution of mathematics (method of proof, style…) and on a useful reminder on definitions viewed in previous lessons. These different kinds of teaching moments enable the teacher to navigate freely between historical and mathematical aims.

The didactic analysis of elaborating processes is helpful for a good description of the wealth of IBST-EHST-ICT sessions. Nonetheless, such a priori study can not be satisfying because it does not take completely into account the way the lesson has been done. The exploratory analysis of different teacher training sessions at the university level has helped to create an analytical framework. The elements so identified can also be used for practice analysis. Now requested by a major part of educators, the evaluation in situ of teachers practice can become a research topic. Some works are already ongoing and there is no doubt that new theoretical frameworks may be invented. Examples below give a way of analyzing the links between science and its history in a classroom situation. Following a traditional distinction, the history of science as a tool and history of science as a goal can appear here as two elements of a unique session.

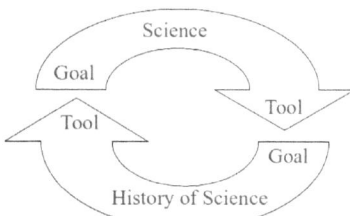

Fig. 2: Science and History of Science

A kind of duality appears and the taking into account of this interactions space opens a new research field for practice analysis. In his classroom, a teacher who combines science and its history constantly juggles in going from a disciplinary domain to the historical domain and then going from the historical domain to the disciplinary one, and so on. The way the teacher manages these swings and all the pedagogical artifices he uses are so much research directions suitable for empirical studies that transcend the traditional academic borders. IBST and ICT are part of the mathematical domain, EHST is the second. The didactic analysis of IBST-EHST-ICT activities can raise elements that make such blends pertinent and give teachers (students, pre-service, in-service) practical professional tools.

References

Barbin, É. (1991). « The reading of original texts: how and why to introduce a. historical perspective », For the learning of mathematics, 11.2, pp.12-14.

Barbin, É. (1997a). « Histoire et enseignement des mathématiques: pourquoi ? Comment ? », Bulletin de l'AMQ (Association Mathématique du Québec), vol.XXXVII, n°1, pp.20-25.

Barbin, É. (1997b). « Sur les relations entre épistémologie, histoire et didactique des mathématiques », Repères-IREM, n°27, pp.63-80.

Barbin, É.; Stehlikova N.; Tzanakis C. (eds) (2008). History ans epistemology in mathematics education, Proceedings of the 5th European summer university ESU5.

De Vittori, T.; Loeuille, H. (2009). « Former des enseignants à l'histoire des sciences: Analyse et enjeux d'une pratique en mathématiques ». Petit x, n°80.

Dorier, J.-L. (2000). « Recherche en Histoire et en Didactique des Mathématiques sur l'Algèbre linéaire – Perspectives théorique sur leurs interactions », Les cahiers du laboratoire Leibniz, 12.

Fauvel, J. ; van Maanen, J. (eds) (2000), History in Mathematics Education, The ICMI Study. Dordrecht/etc: Kluwer Academic Publishers, New ICMI Studies series 6.

Glaubitz, M. (2010). « The use of original sources in the classroom – Empirical research findings », Plenary lecture, 6th European summer university (ESU6) on the history and epistemology inmathematics education, Vienna.

Guedj, M. (2005). « Utiliser des textes historiques dans l'enseignement des sciences physiques en classe de seconde des lycées français: compte-rendu d'innovation », Didaskalia, n°26, mai 2005.

Guedj, M.; Laubé, S.; Savaton, P. (2007). « Éléments de problématiques et de méthodologie pour une didactique de l'épistémologie et de l'histoire des sciences et des techniques (EHST) », IUFM du Nord Pas de Calais. Colloque Théories et expériences dans les didactiques de la géographie et de l'histoire. La question des références pour la recherche et pour la formation.

Jankvist, U.T. (2009). « On empirical research in the field of using history inmathematics education », Revista Latinoamericana de Investigación en Matemática Educativa, 12(1): 67-101.

Jankvist, U.T. (2010) « An empirical study of using history as a 'goal' », Educational studies in mathematics, vol. 74, n°1, pp. 53-74.

Martinand, J.-L. (1993). « Histoire des sciences et didactique de la physique et de la chimie: quelles relations ? », Didaskalia, 2, « Didactique et histoire des sciences », INRP.

Mercier, J.-P. (2004), "Le problème des 5 carrés", Repère – IREM, n°57, pp.47-67.

Raichvarg, D. (1987). « La didactique a-t-elle raison de s'intéresser à l'histoire des sciences ? », Aster, 5, « Didactique et histoire des sciences », INRP.

Rashed, R. (dir.) (1996), Les Mathématiques infinitésimales du IXe au XIe siècle, Vol. I, al-Furquan edition.

REFOREHST (2006), « Histoire des sciences: formations et recherches en IUFM », Tréma, 26.

Rosmorduc, J. (1995). « L'histoire des sciences dans la formation scientifique des maîtres de l'école élémentaire », Didaskalia, 7, INRP.

Siu, M.-K.;Tzanakis, C. (2004). « History of mathematics in classroom teaching – appetizer? main course? or dessert? ». Mediterranean Journal for Research in Mathematics Education, 3(1-2), v–x. Special double issue on the role of the history of mathematics in mathematics education (proceedings from TSG 17 at ICME 10).

Tzanakis, C.; Arcavi, A. (2000). « Integrating history of mathematics in the classroom: an analytic survey ». In: J. Fauvel and J. van Maanen (Eds.), History in Mathematics Education, Chapter 7 (pp. 201–240.). The ICMI Study. Dordrecht: Kluwer Academic Publishers.

Appendix: Geospace figure

Figure Géospace
Numéro de version: 1

Uxyz par rapport à la petite dimension de la fenêtre: 0.1
Rotations de Rxyz: verticale: -23.7325532892 horizontale: 13.90869544059 frontale: -0.09311221414
Repère Rxyz affiché

S point de coordonnées (0,0,3) dans le repère Rxyz
 Dessin de S: non dessiné
C1 cylindre de rayon 1 et d'axe le segment [oS] (unité de longueur Uxyz)
C2 cercle de centre o et de rayon 1 dans le plan oxy (unité Uxyz)
A point libre sur le cercle C2
 Objet libre A, paramètre: 1.01664574958
d droite parallèle à oz passant par A
H point libre sur le cercle C2
 Objet libre H, paramètre: 0.77096946381
 Dessin de H: non dessiné
d2 droite parallèle à oz passant par H
 Dessin de d2: non dessiné
M point libre sur le segment [oS]
 Objet libre M, paramètre: 1
P plan passant par M et parallèle au plan oxy
K point d'intersection de la droite d2 et du plan P
 Dessin de K: non dessiné
B point d'intersection de la droite d et du plan P
Demi-droite [MK)
 Dessin de [MK): non dessiné
L point sur demi-droite [MK), distance à l'origine 3 (unité de longueur Uxyz)
 Dessin de L: non dessiné
P2 polygone régulier d'axe oz, de sommet L, à 4 côtés

Sélection pour trace: B
Parties cachées en pointillé

Plan P isolé par défaut
Fin de la figure

Some secondary school exercises: movement in geometry

15. On fait tourner les formes suivantes autour d'un axe. Quel solide obtient-on dans chaque cas ?

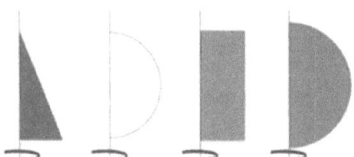

20. On fait tourner le demi-disque ci-contre autour de (DE).
1) Quelle trajectoire va suivre chaque point ?
2) Reproduire le dessin. Placer le centre et donner le rayon de ces trajectoires.

Maths 3e, Bréal, Edition 2008, p.267-268

▷ Dessiner puis découper un demi-disque de rayon 3 cm. Marquer un point M sur son bord et un point T à l'intérieur. Scotcher le demi-disque sur un crayon et faire tourner ce dernier rapidement entre les mains. Lequel des trois solides ci-dessous peut-on alors apercevoir ? Quelles lignes décrivent M et T ?

Maths 3e, Bréal, Edition 2008, p.257

PART II

Inquiry based mathematics teaching and the history of mathematics in the English curriculum

Snezana Lawrence

Bath Spa University, England; s.lawrence2@bathspa.ac.uk; snezana@mathsisgoodforyou.com

ABSTRACT: The project at the centre of this paper involved working with original sources from the history of mathematics, and the video conferencing between different schools to support students in their research projects. It is suggested that the secondary curriculum in mathematics offers possibilities to develop resources which can support and lead students to conceptually populate and create their own learning landscape which is primarily based on inquiry-based learning practices. Issues for teacher development and collaborative learning and teaching make part of such practices, and the role the history of mathematics can play in this context is examined.

The organisation of inquiry-led learning project

The described project began in September of 2006 in a group of schools in the South East of England,[1] and was completed in the Summer of 2009 with a teacher interchange between two schools across the Atlantic Ocean – one school being in England and another in the United States. The first phase of the project began as an attempt to answer a question about what kind of model of continuing professional development would engage teachers on a longer term basis. Our suggestion was that collaborative research and teaching, having as a focus the history of mathematics, could enrich the experience both of teachers and of their pupils, and could sustain collaboration over prolonged periods of time.

A short background to this is needed. Professor Adrian Smith was commissioned by the Royal Society of Great Britain and the ACME (Advisory Committee on Mathematics Education) in 2004 to review[2] the teaching and learning of mathematics in secondary schools in the UK. One of the outcomes of a number of his recommendations was the founding of the National Centre for Excellence in the Teaching of Mathematics (NCETM). The consensus was at the same time established amongst the mathematics education community in the UK[3], that the collaboration, networking and awareness of a variety of practices and resources, and the self-study and research in peer groups, are the most valid forms of continuing professional development in mathematics education. To support the development of such a model of practice, the NCETM, immediately upon their foundation, issued a call for proposals for small-scale projects which fit the criteria of collaboration and research, and our project, entitled at the time 'History of Mathematics and Collaborative Teaching Practice', was one which was supported through the first round of grants between 2006/7 and 2007/8 academic years.

At the end of the second year of the project, it became clear that the pupils who participated were able to take on certain tasks of leading the learning in groups themselves. A lot has been written on the dangers of 'minimal guidance' or inquiry-based learning and teaching,[4] but it has been acknowledged even by the hard-core instructional educationalists that "the advantage of guidance begins to recede… when learners have sufficiently high prior knowledge to provide 'internal' guidance".[5] Hence the second part of our project was conceived to be led by those pupils who have 'high prior knowledge', as much as their interests to lead this inquiry could be developed individually, and within a framework limited and defined by their teachers and the choice of a mathematical topic.

The project was thus organised by two teachers, one working in a school in England and one in the US, who constructed an inquiry-led course for pupils to study the History and Theory of Polyhedra, and base their enquiry on the historical advancement of the study area.[6] Two groups were almost entirely disparate – the only thing in common being the pupils' high ability in mathematics.[7] Pupils were invited to join the project in two ways: pupils from England were given an invitation to engage with the project in their spare time during the year, but with a view of having a two-week course at the end of the academic year. During these two weeks an intense course was organised by both teachers (the US teacher joined the UK pupils and their teacher) and the students were withdrawn from other subjects for the duration of that final sequence of the course. Pupils from the UK were between the ages of 12 and 18. At the beginning of this part of the project, around 37 pupils opted to work in the group which they called 'The Langton Institute for Young Mathematicians',[8] but only 23 attended the course in an intensive and productive way throughout the academic year.

Pupils in the US however were given the option to take the course in the History and Theory of Polyhedra as part of their yearly academic 'diet'. These were pupils between the ages 16-18, and the group was fairly small compared to the English 'set' – at any time there were no more than 5 students in the American group. This paper traces the project mainly on the basis of observation of the UK students' progress and refers to the American group only as a 'evaluative' peer review group.

The system of instruction was organised as a repetitive cycle:

- During the first lesson in a cycle students were given 'instructive' lesson, during which they had learnt something new from their teacher
- Second lesson was based on discussions around the questions students had wanted answers to following the instructional lesson – they would then collaboratively agree upon the research that needed to be done and divided tasks between themselves
- Between second and third lesson students would have approximately a week of research time to complete their work either in the library, mathematics lab in school, or at home
- Third lesson was organised around a video-conference during which students on both continents would communicate their findings and pose further questions to each other and their teachers.

Teachers followed students' progress and modelled their 'instructional' lessons to suit this progress. Communication between teachers continued between the lessons, in order to plan a further course of action. Although planning for the course proved to be research intensive, teachers also drew on the expertise of local academics and professional mathematicians, who made visits to the schools to discuss questions of interest with the pupils. One such lesson, for example, was the application of the Theory of Polyhedra to computer programming and security,[9] and another was based on the narrative experiences of a professional mathematician and his ability to transfer his knowledge of mathematics to optimisation problems in marine engineering.[10]

Change of the curriculum

The project began when the changes to the curriculum meant that the history of mathematics became an entitlement for every child in the UK[11], stressing in particular the need for all students to recognise the 'rich historical and cultural roots of mathematics':

> Mathematics has a rich and fascinating history and has been developed across the world to solve problems and for its own sake. Students should learn about problems from the past that led to the development of particular areas of mathematics, appreciate that pure mathematical findings sometimes precede practical applications, and understand that mathematics continues to develop and evolve.[12]

A little less than two years after the completion of the project however, the National Curriculum in the UK is being revised again by the new coalition government,[13] having the main remit 'commitment to give schools greater freedom over the curriculum'. As the curriculum changes and more 'freedom' is given to teachers, questions are being raised about how will teachers cope with this newly acquired freedom. Whilst changes to the curriculum always bring certain concerns to teachers and other 'deliverers' of the curriculum, the project described here could well be taken as a particular model of both development of the curriculum and a model of leading the learning in a way in which gifted and talented pupils may truly engage in learning alongside their teachers and develop a learning landscape that all pupils can access in some ways.

A crucial question raised at this point however, is that of expertise and the soundness of our claim that students themselves are able to work alongside their teachers thus leading the inquiry.

> Most learners of all ages know how to construct knowledge when given adequate information and there is no evidence that presenting them with partial information enhances their ability to construct a representation more than giving them full information.[14]

To deconstruct this statement we ask whether there is such an instruction in which all information is possible to be given to a student about the topic that is being taught. Of course, this is not the case – there is no teaching method which would convey every possible meaning that a mathematical concept contains, or even every possible connection between that topic and the other areas of mathematics. Even in the most stringently constructed curriculum, when presenting a mathematical concept, the teacher has limited time, resources and knowledge which he or she can impart to students through the 'provision' of that information. The attitude however, of presenting mathematics syllabus-linked information as the totality of available knowledge and putting a learner in a position where they should 'accept' that knowledge, raises the questions not only of the nature of possible knowledge but the nature in particular of mathematical knowledge, apart from the very obvious motivational forces in the learning of mathematics. Is all the possible knowledge in the hands of teachers who should hand that to the new generations? It would be nice perhaps to think that this is so, even having in mind the impossibility of such a situation, but recent studies[15] suggests that we should be more practical when faced with the realities of teaching mathematics. Add to this the fact that the curriculum seems to be constantly changing (no bad thing), as well as the enormity of the mathematical knowledge that is available and being developed in the modern world, and we face the situation that we already have in most classrooms – where mathematics is regarded as a monolithic slab of knowledge too heavy to be taken up by teenagers to deconstruct, examine, and enjoy, because teachers construct their teaching around the premise that they alone are bearers of this knowledge.[16] Confining the learning of mathematics to what the teacher knows, seen from this angle, is not productive for either pupils or their teachers.

Returning to the project, we now examine its findings to plot the possible route through which students can feel enabled to take on mathematics as a creative discipline, in which new questions can be asked, and which lead to collaborative learning not only between students but with teachers (and between teachers), and are led by students' interests.

As mentioned above, the project consisted of two parts: first dedicated to introduction of collaborative research and learning, in which both teachers and pupils gained expertise relevant to the subject matter and the focus – historical study of mathematical discoveries – and second dedicated to freeing the constraints placed on pupils through the curriculum too narrowly defined, in which pupils were encouraged to pose questions and lead enquiry of the group.

It became apparent, in the second, inquiry-led learning part of the project, that a plethora of possible learning 'routes' was available, but three major routes were defined by the groups of pupils who organised themselves to pursue them:

1. Learning through problem solving, leading to the learning of mathematical facts and techniques.
2. Studying mathematical proof.
3. Representation, generalisation and abstraction through the study of mathematics in context[17] was the work of the third group, with sub-

groups/individual students doing work leading to a greater understanding of polyhedra through the:
   a. generalisation and categorisation of objects, leading to abstraction of mathematical concepts
   b. cultural mathematics and the applications of mathematics.

Elaborating on the three main routes

1. Problem solving in constructive geometry – the properties, history, and building of models of Archimedean solids, and solids derived from these.

Fig. 1: Constructing the rhombic hexacontahedron from net (yellow model).

This group began by investigating the history and theory of polyhedra in general, learning about properties of Platonic and Archimedean solids, and constructing mathematical models during which those properties were further investigated. The group chose to investigate in more depth the rhombic hexecontahedron, a 60-faced polyhedron which can be obtained by stellating rhombic triacontahedron, a convex polyhedron with 30 faces. The latter is an Archimedean dual solid, or Catalan solid,[18] and is a dual of icosidodecahedron. The chain link of names provided students with opportunities to chart the development of the theory of polyhedra, as well as mathematical properties of the same. The outcome of this project was the construction of the rhombic hexacontahedron first as a solid made from a net (Fig. 1) and secondly as a sculpture made from slide-togethers (Fig. 2).

Fig. 2: The model made from 'slide-togethers'.

2. History of proof in mathematics

This group started their work by studying and digitising an 18th century English school-edition of Euclid's *Elements*[19] (Fig. 3) then studying and comparing the text with their current textbook, and its content on geometric proof. Furthermore, they looked at the validity of one proof: angles in a triangle add to 180° (Proposition I.32), related this to propositions 1.27-1.29, and investigated the sum of angles in spherical geometry. Subsequently this groups' main work consisted of investigating the history of the development of Euclidean and non-Euclidean geometries. The discussions of this group centred on a few issues:

- What is mathematical proof – how do we define it and how did other generations define it?

- How does the learning of mathematics change over time? What can we see from an old textbook and how can we compare it to the new one?

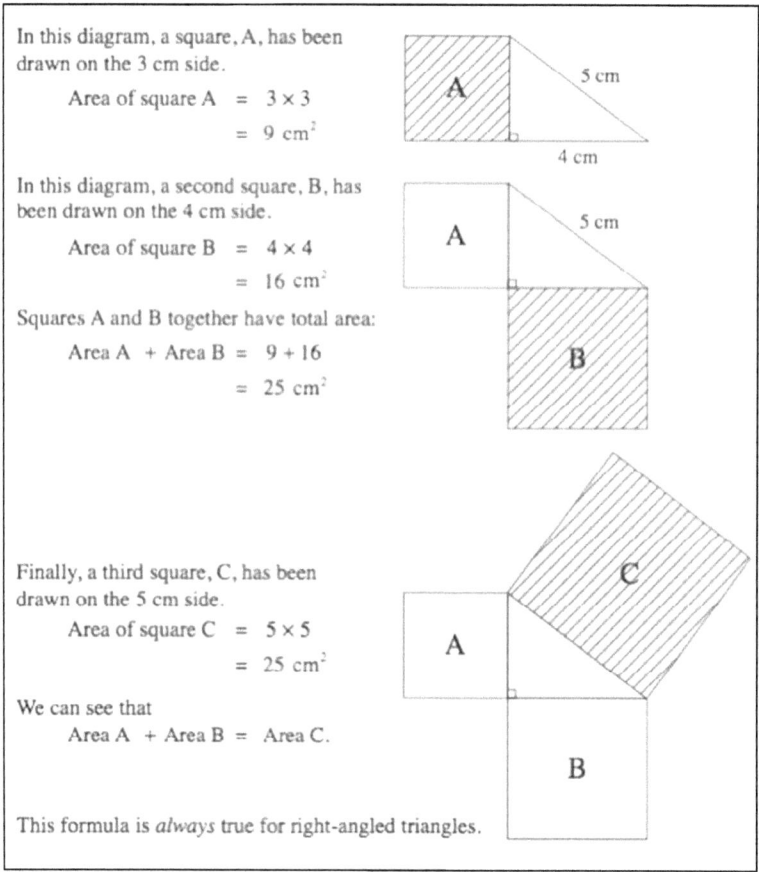

Fig. 3a: MEP (Mathematics Enhancement Project, http://www.cimt.plymouth.ac.uk/projects/mepres/book8/bk8_3.pdf)

An example of the answer to the question about the nature of mathematical proof was given by students who compared the proofs given in Simson (1756) and MEP (1996).

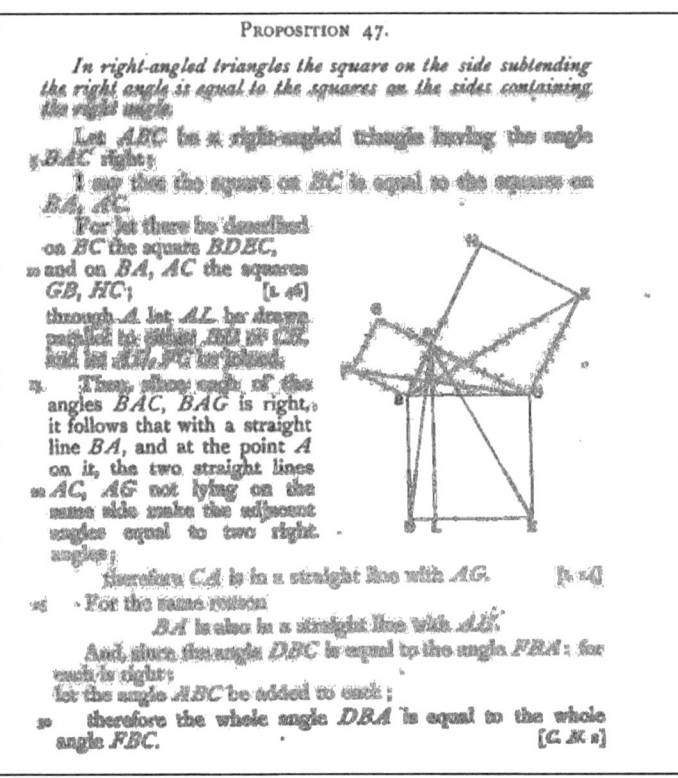

Fig. 3b: The proof as given by Simson 1756.

In the view of the students, Simson's proof was much more difficult to understand than that given in MEP (1996). The students first did not see the difference other than in the different levels of 'difficulty' between the two proofs, but after a discussion came to distinguish that MEP 'proof' had not actually proved the theorem, but demonstrated how it works, whilst Simson's proof used deductive reasoning to show that this theorem always worked for right angled-triangles. Whilst this was an important discovery for pupils: understanding that 'proof' in fact covers the statement in all cases, and that demonstration merely shows how a theorem 'works', teacher also gained an insight into the process of how the rigour of mathematical proof has (sadly) changed so that what is called 'proof' is not in fact covered by the curriculum material recommended for the classroom use, but merely refers to demonstrations given to show how the theorem works in several cases.

An interesting and entirely different aspect of working with original sources gave an additional 'lesson' for the teachers: the pupils engaged with a non-mathematical content – scribbles in the margins of the examined book led to discussions raging from the banal 'he obviously didn't like his teacher much' to learning about important mathematical and historical facts about Fermat and his 'marginal' writing.

Representation, generalisation and abstraction

This group used presentations to share their findings, one of which for example consisted of the categorisation of polyhedra as they appeared in art during the Italian Renaissance (Fig. 4). Students in this group could perhaps be best described as 'mathematical poets' – their main aim being the appreciation of, and search for, beauty in mathematics and mathematical artefacts.

Although in 'everyday' lessons we often forget about this aspect of mathematics, this group was in fact central in many ways to keeping the project going, and giving the whole UK group a certain social sense of belonging, a social but also intellectual raison d'être, which will be further elaborated upon. This group was also most active in setting up and developing technical infrastructure for the project, from video-conferencing to setting up an on-line chat room for UK and US students to interchange information about their work. The chat-room was developed and used by pupils to communicate with their colleagues in US on practical matters: for example how did they interpret the images of polyhedra, where they found them, and how did the images of polyhedra develop throughout the history.

One English group of pupils in particular did a research on the representation of polyhedra in Art during the Renaissance. This was very much appreciated by a teacher of history (in the English school where the project took place), who was very much amazed at the information pupils learnt from the history through the project they were doing in mathematics. Renaissance period is not covered by the English History Curriculum, so the fact that pupils learnt about the period itself was something unexpected to this history teacher.

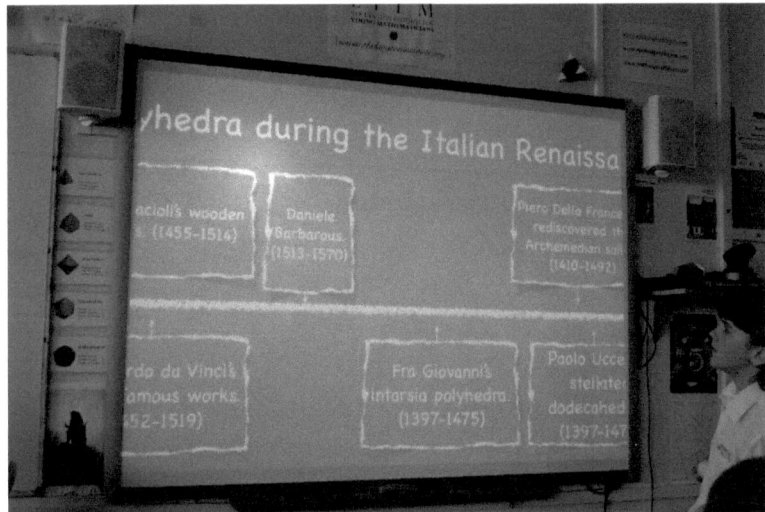

Fig. 4: Year 7 pupil (age 12) presents his generalisation and categorisation of polyhedra that appeared in art in the Italian Renaissance.

The role of the history of mathematics in inquiry-led learning

There are several categories mentioned above which give examples and note possibilities for introducing the history of mathematics and its role in an inquiry-led learning environment. We will now further examine them in order to categorise the role the history of mathematics can play in inquiry-led learning.

It is important to state at this stage that the history of mathematics was not introduced as a goal[20] in this study but as an element to create a necessary milieu in which it is possible to create a community of enquiry and initiate students into the discipline in a way that an apprentice may be introduced to a profession. The most important role of the history of mathematics in this respect can be seen in two ways, which we describe as follows.

Firstly, using the historical development of the study of subject matter to narrow down the inquiry pathways for both students and teachers in order to give structure to all the possible (innumerable) routes of enquiry. The problem that the history of mathematics aided in solving in this respect was its role in putting in context the nature and number of questions that pupils may have in relation to mathematical knowledge, and being able to develop questions and guide pupils through a historically validated landscape of mathematical discoveries related to the general topic (Polyhedra). These 'questions' include all the other mentioned routes of enquiry explained above. While pupils worked in groups and dedicated most of their project time to one or two aspects investigating the theory of polyhedra, from 'problem solving', 'history of proof' to 'representation of polyhedra', they were nevertheless aware of other groups' findings. This meant that, for the first time in their experience of learning mathematics, pupils could choose the pathway through which they would learn the subject. Additionally they felt in 'control' of their learning and considered themselves as 'leaders' in their particular way of investigation as they had to explain their findings to pupils in other groups.

We were aware of the connotations that some current educationalists have against the teaching of 'the discipline by inquiry':

> The major fallacy of this rationale is that it makes no distinction between the behaviours and methods of a researcher who is an expert practicing a profession and those students who are new to the discipline and who are, thus essentially novices.[21]

Secondly, at the same time we emphasised that mathematics as a discipline cannot be learnt without an ability to engage with an inquiry in the same way perhaps that painting cannot be learnt by simply mixing colours, not withstanding the expertise of an established painter and his ability to teach a novice in his care about the creative processes and not only techniques.

> In this respect we tried to 'push' the learning process from that of 'teaching a discipline by inquiry'[22] to that of learning to 'do' mathematics in a community of practice.

Observing the prevalent practice in which mathematics qualifications have become the main aim of mathematics education, we aimed to introduce the study of mathematics as an intellectual discipline and one which may become an identifiable and valid choice of profession for our students in the future. In other words, we tried to support our pupils in pursuing their study of mathematics in a way that their competence in the subject, for examination purposes, will be secondary to their appreciation, enjoyment, and development of their ability in learning of mathematics, which we considered to be a primary aim of mathematics education and our guiding 'light' in this project. As Watson (2008) has said:

> ...it is becoming more and more the case that, in an effort to ensure that more students can gain school mathematics qualifications, the difficult shifts which would have to be made for school mathematics to be a subset of the discipline of mathematics are being edited out of mathematics as a school subject, rather than edited in as the discipline itself becomes more complex, more post-modern and less certain.[23]

In other words, the primary aim of our educational system seems to be developing competencies related to passing the examinations rather than competencies related to the actual study of mathematics.

It is perhaps, when mentioning this 'less certain' aspect of modern mathematics that we should look at the important role of the history of mathematics in inquiry-led learning of the subject. It has long been established in previous studies of the role of the history of mathematics in education,[24] that the history of mathematics gives possibilities to introduce students to a discipline in a contextual way.

Learning about the 'old' masters, trying to recreate the old and famous pieces of their work, and having a narrative to set this work in context, also gave students an insight that discovery, and not perhaps knowing where it can lead to, asking questions and perhaps not being able to answer them (within a span of centuries, rather than in a few minutes within a mathematics lesson as is usually expected of them) is part of 'doing' mathematics. In particular the realisation that this may be so, came through investigations on the history of proof, discussions on Euclidean and non-Euclidean geometries, and famous unsolved problems in mathematics, leading students to be empowered to search further without necessarily getting the quick answer or an outcome that can be presented to an examination board.

Far too often we seem keen to avoid having students not complete their 'work' – the exercise diet that we expose them to in the mathematics curriculum seems to be used to fill their exercise books with evidence that that 'work' is being done. While it is quite certain that 'a state of conflict may easily lead to frustration and fragility'[25], not giving the students the full details of how to deal with the conflict (of not knowing the right answer) in mathematics and allowing them to find the routes of learning that will lead them to better understanding the cognitive 'conflict' that they may be experiencing, should also be part of the process of learning the subject. We tried to teach this process of conflict resolution 'through the construction of knowledge structures which are consistent with domain principles'[26], having the history of mathematics as a prime guiding principle to show how others have done similar things in the past.

Other factors to take into account

Certain constraints were identified during the project which all needs to be taken account of in case a similar approach is tried or adopted.

Firstly we found that teachers had to have a desire to engage with learning themselves. As they also teach 'traditional' courses in addition to an inquiry-led course like that described in this paper, they had to accept that a role which clearly defines them as an expert who 'delivers' the curriculum is interchangeable with that of a facilitator of learning who tries to find answers to questions either by him or herself or find another expert who may be able to help the students in their research.[27] Teachers in this project reported that the greatest benefit for them was in learning how to construct such an environment of inquiry.

Very practical concerns about the timetabling (giving the students time off other lessons to participate in the lessons and activities, including the constraints related to the difference in time-zones between the two schools and the video-conferencing which was integral part of the project), meant that senior management, in particular the heads of schools, had to give full support to such an undertaking. Fortunately both schools had inspiring and visionary people at the top, who were not only glad to support the project, but in their various ways engaged with students in a dialogue during, and following the completion, of the project.

The students who participated were all gifted and talented kids –and extra work was something that they were used to, and their enthusiasm and independence flourished during and after the project. Some continue communicating their latest research two years after the completion of the project. But what happens if we try to engage those less able in a learning environment such as this?

> One outstanding feature of UK mathematics teaching which endures through all changes in school organisation, assessment systems, policy and imposed 'normal' practice is the segregation of students from each other in order to be treated to a different curriculum, with different expectations, and that these differences are discussed in terms of fixed deficiencies for the most vulnerable and potentially troublesome adolescents.[28]

In this respect we did not make much progress, as we had time constraints to engage only those pupils who opted to do so themselves. The real benefit though, could be further explored for those pupils who believe that they are not 'good at maths' and in encouraging them to engage in 'doing' mathematics in the same way – by asking questions and leading the learning themselves. This yet remains to be done.

Conclusions and new questions

Whilst we engaged with students of ages ranging 12-18, the project showed us that

1. Students who persisted with the project to the very end did so because they constructed not only their learning 'pathway' led by their interests, but also because they were part of a collegial group of similarly minded students. The historical narrative of the development of mathematics in this respect had a positive influence on their own sense of belonging to a tradition as well as to a group of similarly minded people.
2. The age of students did not have an influence on their confidence to engage with the project. In this respect the view that what Young and Lucas (1999) describe as a

   …typical shifts in the pedagogy of most subjects as learners grow older… include shifts from teacher-centred to learner-centred organisation; teacher-authority to learner-enquiry; insular to connected knowledge. These shifts parallel what is being said about the learning of mathematics, in that the inherent authority of mathematics, the need for problem-solving approaches, and the need for discussion match a shift to more autonomous organisation. Sadly, in mathematics some of the opposite shifts tend to be the case, in that the demands of abstract work lead to an increased focus on the teacher as authority…[29]

we did not find to be entirely valid in our situation.

Finally, in relation to the point (2), we found that the confidence to engage with an inquiry-based, autonomous way of learning mathematics, was more a function of a narrative both of their social group and the narrative of mathematical development they identified with, rather than students' age. This leaves us in a position to recommend an approach as described in this paper to be tried in settings across the age range of secondary mathematics education.

References

Behr, M. and Harel, G. (1995) *'Students' errors, misconceptions, and conflict in application of procedures'*, in Focus on Learning Problems in Mathematics 12 (3/4), 75-85.

Bell, A.W. (1979) "The Learning of Process Aspects of Mathematics", *Educational Studies in Mathematics*, 10:3, 361-387.

Davis, and Simmt, (2006) "Mathematics for Teaching: An Ongoing Investigation of the Mathematics That Teachers (Need to) Know", *Educational Studies in Mathematics*, 61:3, 293-319.

Fauvel, J. and Swetz, F. (1995) *Learn from the masters!* Mathematical Association of America.

Jankvist, U. (2009) "A categorization of the 'whys' and 'hows' of using history in mathematics education", *Educational Studies in Mathematics*, 71:3, 235-261.

Kirschner, Sweller, and Clark, (2006) "Why Minimal Guidance During Instruction Does Not work: An Analysis of the Failure of Constructivist, Discovery, Problem-Based, Experiential, and Inquiry-Based Teaching", *Educational Psychologist*, 41:2, 75-86.

Lawrence, S. (2008) *History of Mathematics Making Its Way Through the Teacher Networks*, Paper presented at of ICME11, 7th July 2008, Mexico City, Mexico.

Lawrence, S. (2009) What works in the Classroom – Project on the History of Mathematics and the Collaborative Teaching Practice. Paper presented at CERME 6, January 28th, 2009, Lyon France.

MEP Mathematics Enhancement Project (1996-2010).
http://www.cimt.plymouth.ac.uk/projects/mepres/book8/bk8_3.pdf.

Naeve A., (1997). *The Garden of Knowledge as a Knowledge Manifold - A Conceptual Framework for Computer Supported Subjective Education*, CID-17, TRITA-NA-D9708, KTH, Stockholm, accessed September 1st 2008,
http://cid.nada.kth.se/sv/pdf/cid_17.pdf.

Simson R. (1756) *Euclid's Elements*. Glasgow.

Smith A. (2004) Inquiry into post-14 mathematics education,
<<www.tda.gov.uk/upload/resources/ pdf/m/mathsinquiry_finalreport.pdf>>.

Watson, A (2004) "Red Herrings: Post-14 'Best' Mathematics Teaching and Curricula", *British Journal of Educational Studies*, 52:4, 359-376.

Watson, A. (2008) School mathematics as a special kind of mathematics. In *For the Learning of Mathematics*. 28(3) 3-8.

Zaslavsky, O. (2005) "Seizing the Opportunity to Create Uncertainty in Learning Mathematics", *Educational Studies in Mathematics*, 60:3, 297-321.

---

[1] This part of the project has already been described fully in Lawrence (2008) and (2009).

[2] The UK Government commissioned a review into the teaching of mathematics into post-14 mathematics education, the results of which were summarized by professor Adrian Smith in Smith (2004).

[3] In particular see the aims of the NCETM (National Centre for Excellence in the Teaching of Mathematics at https://www.ncetm.org.uk/ncetm/about. Accessed 16th October 2010.

[4] Kirschner et al. (2006).

[5] Kirschner, P, Sweller, J, Clark, R (2006), 80.

[6] See http://www.vincematsko.com/

[7] Both schools catered for gifted and talented pupils, and the pupils invited to join the course were from the top ability sets, therefore falling into the category of the highest 5% ability level in mathematics.

[8] The 'Institute' was based at the Langton School for Boys, Canterbury, UK.

[9] Delivered by Dr Andrew King from the University of Kent.

[10] Delivered by Mr Nira Chamberlain, from the Institute of Mathematics and Its Applications.

[11] The National Curriculum of 2008.

[12] Page 4 of the QCA Mathematics Curriculum, accessed 20th March 2008, <<http://curriculum.qca.org.uk/subjects/mathematics/keystage3/index.aspx>>.

[13] The new government took office on 11th May 2010.

[14] Kirschner, P, Sweller, J, Clark, R (2006), 78.

[15] For example, Davis, B and Simmt, E (2006).

[16] It may not then come as a surprise to hear that the perception that many young people in Britain have of mathematics is that it is "boring and irrelevant" as a consequence (Smith, 2004, 2).

[17] Bell, AW (1979), 362.

[18] Catalan solid, or Archimedean dual, is a dual polyhedron to an Archimedean solid.

[19] Robert Simson edition, 1756, Glasgow.

[20] See Jankvist, U (2009).

[21] Kirschner, P, Sweller, J, Clark, R (2006), 79.

[22] Kirschner, P, Sweller, J, Clark, R (2006), 78.

[23] Watson (2008), 4.

[24] Fauvel, J and Swetz, F (1995), to mention one of the very many.

[25] Zaslavsky (2005), 318.

[26] See Behr and Harel, 1995:83.

[27] See Naeve (1997).
[28] Watson, (2004), 369.
[29] Watson (2004), 368.

ICT and History of mathematics in the case of IBST

Olivier Bruneau[*], Sylvain Laubé[+] & Thomas de Vittori[#]

[*] *Université de Lorraine, Laboratoire d'Histoire des Sciences et de Philosophie - Archives Henri Poincaré, UMR 7117, Nancy, F-54000, France; olivier.bruneau@univ-lorraine.fr*
[+] *Centre François Viète (EA 1161), Université de Brest, France; sylvain.laube@univ-brest.fr*
[#] *Laboratoire de Mathématiques de Lens, Université d'Artois, France; thomas.devittori@euler.univ-artois.fr*

ABSTRACT: Following a European project on the relationships between inquiry based science teaching (IBST), internet and communication tools (ICT) and a third topic, history of science, this article deals with some of the new questions raised by the recent introduction of epistemology, history of science and technology (EHST) in the school curricula. The definition of the inquiry and the role that can be played by epistemological thinking in science education are especially analysed in this text. Based on the analysis of historical examples and many on-line resources, as a first result, a new framework for the elaboration of new resources in IBST, ICT and EHST is given and some opening questions have been pointed out.

Context: IBST in Europe

All over Europe, the lack of student interest in science or in scientific careers has been noticed for years. This situation has led the European Community to launch a call for research projects in science education (the FP7 Science in Society program) and the publication of the Rocard Report about Science Education[1]. One of the main recommendations was to promote the advancing of teaching methods toward Inquiry Based Science Teaching (IBST) and the request for international comparisons. Engaged in 2008[2], the European research project "Mind the Gap" was one of the answers. "Mind the Gap" was a large didactic program in which about 15 European universities worked on inquiry based science teaching in order to develop useful scientific tools within this topic. Our research group of historians (named *PaHST*[3]) in the University of Brest join this program. In a first meaning, IBST consists in learning based on an open problem in which the student has to propose experiment or use instruments in order to find a solution. Such inquiry based teaching can be divided into different aims: an understanding of the articulation between empirical evidences and concepts (e. g. testing hypothesis, modelling, results evaluation), the practice of hands-on activities comprising an informational quest or not (experimentations), the introduction to a specific scientific language (argumentation, debate), the enhancement of students' autonomy…Mind the Gap was a place wherein to think about the interaction between IBST, history of science and technology (HST) and Internet and communication tools (ICT). In this article, we would like to render an account of one of these possible interactions and to show thereafter how historical on-line resources can constitute well-adapted references of authentic problems.

IBST in Mathematics: a historical answer

According to Pr. Martin Andler (University of Versailles – France), a contemporary mathematical research activity comprises: 45% devoted to observation, 45% to experimentand only 10% to demonstration[4]. In the field of mathematics learning, the inquiry-based style often claims to have been inspired by the "scholar at work". In mathematics as in the other fields of science, inquiry takes up a large part of the research process. There is no doubt about the fact that the job has changed throughout the ages. However, by rendering an account of these changes, the epistemology and history of mathematics can help to explain what this inquiry could be.

First example

In ancient times in favouring results to reasoning mathematicians rarely expressed themselves on their relationships to the experiments. However when they did so, they gave us the opportunity to see the complexity of the links between theory and the use of technical instruments. The history of science assures us that: mathematical theories never emerge from nothingness. The scientist describes, builds and explores multiple examples before proposing an analysis or a system. On this subject, the work of the Arabic scholar al-Sijzi is a model of such a process. Ahmad ibn Muhammad ibn Adb al-Jalil al-Sijzi was born and lived in Iran. Son of a mathematician, he worked between 969 and 998 and he wrote exclusively books on geometry. In all, he has written approximately fifty treatises and lots of letters to his contemporaries. Al-Sijzi was working within a very specific scientific context. Since the ninth century, the development of algebra in the Arabic world has created new types of questions on the fundaments of this field. For instance, the point-by-point construction of conics has been well known since Antiquity (see the Apollonius' book entitled the *Conics*, for example), and that method is efficient enough for the analysis of the main properties of those curves. But during the ninth century, as the major part of algebraic equations studied during that period can be solved by intersecting conics curves (ellipsis, parabola, and hyperbola), the necessary taking into account of these intersections creates new difficulties. Indeed this possibility is based on the continuity of the different curves which is difficult to "prove". The solution that has therefore been chosen is to associate the curve with a tool that enables a real construction. As the ruler and the compass allow straight lines or circles to be drawn and so justify their continuity, a new tool had to be invented to draw all the conics. Not only interesting from a mathematical point of view, such a technical development is also useful in technological areas such as the construction of astrolabes and sundials where conics are essential.

Following his predecessors (Banu Musa, Ibrahim ibn Sinan...) from whom he quoted in a precedent book on the description of the conic sections, al-Sijzi engages himself too in a treatise specifically on the *Construction of the perfect compass which is the compass of the cone*[5]. In this book, he studies a new tool, made-up a while ago by al-Qūhi: the perfect compass. As he announces in the first pages of his treatise, al-Sijzi wants to "build a compass which enables him to obtain the three sections of the cone." He first notes that all

147

the conics can be obtained from the right cone (depending on the position of the cutting plane), and afterwards he proposes three possible structures for the perfect compass. The beginning of the study is technical, "We must now show how to shape a compass by which we can trace these sections. Shaping a stalk or AB. We put on the top tube, or NA. We link another tube to the end of it. [...]", and these instructions should enable the reader to build such a compass. But for al-Sijzi, the aim of his work on the perfect compass is not only to draw conics. The end of the text shows that this compass is also a theoretical tool and a tool for the discovery of new concepts. Al-Sijzi explains that the link between the circle and the ellipsis is quite obvious. Indeed, the construction of the ellipsis by orthogonal affinity and the formula for the area are both well known. But what are the links between the circle and the parabola or the hyperbola? Now oriented towards the exploration and the solving of new problems, the practical tool becomes an instrument of discovery and as stated by al-Sijzi: "I always thought that there was a relationship between these two figures and the circle and their similarities and tried to get it but the knowledge of this has only become possible to me once I had learned how to turn the perfect compass following the positions of the plans."

In this example, the comings and goings between theory and practice appear clearly. Confronted with the theoretical problem of the continuity of curves, the scientist suggests the use of a new instrument. The experimentation with this instrument creates new theoretical results that create new questions and so on and so forth. In this text, al-Sijzi clarifies the role of mathematical instruments. They are objects as much as models and this dual status facilitates the theory-experiment passage.

Another example

Pierre de Fermat (1601-1665), the French lawyer who works during his free time on mathematics, is well known and his name is associated with the famous theorem which was only demonstrated in the 1990's. Nowadays, we can find[6] some parts of his mathematical works on the theory of numbers on the Web. Even if he has not published any treatise on this subject, some elements can be found in quotations of his Diophantus' book and included in his correspondence especially with Marin Mersenne, Huygens or Carcavi. In a letter Fermat sent around august 1659 to Pierre de Carcavi (1600-1684), another amateur French mathematician, he tried to demonstrate some properties with, according to him, a new kind of demonstration he called "la descente infinite ou indefinie". This kind of demonstration is also used in the margin of his Diophantus' volume in order to prove that the area of rectangular triangle with integer sides (i.e. ones which are measured by integers) cannot be a square. Up to 1659, Fermat only uses this method to prove some negative results. He proposes a positive result, for the first time, in his letter to Carcavi: any prime number such as $4k+1$ is the sum of two squares, and this form is unique. For instance, 5^2+1, $13=3^2+2^2$, ... The demonstration is based on a reduction *per absurdum*. For instance, to prove the first above proposition, he presupposes that this kind of triangle exists. Then he shows that if it is true, we can find another triangle with shorter sides which validates the assertion, and so on. Since the sides must be integers he arrives at a contradiction through the method of infinite descent. It is quite easy to prove negative assertions, but it's harder to show positive assertions. In his letter, Fermat announces that he only demonstrated that any prime number which could be written as $4k+1$ could be decomposed as the sum of two squares of integers and that this decomposition was unique but, unfortunately he did not give any demonstration. The first proof of this assertion seems to appear in Leonard Euler's (1707-1783) works under the Latin title of *Demonstratio theorematis Fermantiani omnem numerum primum formae 4n+1 esse summam duorum quadratorum* [Demonstration of the Fermat's theorem "All prime number like $4n+1$ is sum of two squares"][7], in which he strictly follows the Fermat's method.

The method of "infinite descent" has profoundly renewed the theory of numbers. Even if, as usual, Fermat does not demonstrate what he announces, he suggests to the other mathematicians that they can easily find the demonstration of the properties he gives. For instance, in the above-mentioned letter, he does not give any demonstration, however, he asks his contemporaries to do so: "je serai bien aise que les Pascal et les Roberval et tant d'autres savants la cherchent sur mon indication" (*op. cit.* p. 432). In this letter, he shows two important aspects of mathematics. First, some methods of proof are not able to solve a problem and newer ones must be invented and then reused to solve other mathematical enigmas. Creativity can go through a new method and a new approach. Secondly, this letter suggests to the other mathematicians to deal with those kinds of problems. This demonstrates how mathematical progress can be transmitted.

Many types of inquiry

The reading of the ancient texts is always interesting. Both examples presented above are only a small part of the historical resources dealing with inquiry, but they are still meaningful. What the historical approach shows is that *inquiry* cannot be reduced to a single aspect of the research process. Depending on the situation, each scientist engages himself in a different type of inquiry. A theoretical question does not require the same method as an experimental one, etc. IBST is clearly inspired by the work of the scientist as a professional. Nonetheless, the way this one has been understood is sometimes a bit caricature. The history of science can prevent us from simplifying to such an extent and can restore the wealth of the research process.

History of mathematics and on-line resources

Nowadays, on the web, anyone can find many websites, pages or documents related to the epistemology, history of science and technology (EHST). Some sites like "Internet Resources for History of Science and Technology"[8] have even been created to help users to find the right resources. More and more primary sources are also available on "Google books" or on the French project, "Gallica"[9]. It is relatively easy to read and to study ancient texts on science and technology. But in order to avoid mistakes or misinterpretations, it has been demonstrated that the texts must be contextualized. This is one of the main results of the research on the history of science. Now, due to the development of the new technology and the easiness of the on-line publication, some websites are devoted to IBST but they do not include any historical aspects. Is it possible to conciliate the two aspects: IBST with a historical approach? In the Mind the Gap research program, the study of the relationships between inquiry and the history of science has shown that science and investigation are very close in different ways. In every scientific field, the scholars have to elaborate a way of questioning their object. In each case, new instruments should be built, and new experiments should be elaborated in order to be able to ask the right questions and finally to answer them. This first part of scientific activity is very similar to a second one which is dealing with theories and models. When their pertinence has been proved, the new concepts are applied to other situations or fields and they become a part of common knowledge. These newly born theories have still to be discussed and they constitute a third part of scientific activity. Communication between scholars often needs to create a suitable language and is essential in the building process of knowledge. An on-line publication should take into account all this wealth that gives many opportunities to the enlightening of historical documents and in making it suitable for IBST in multiple ways.

Digital document for EHST

According to Michael Shepherd and Livia Polanyi[10] quoted by Ioannis Kanellos[11], the genre of digital documents can be characterized through three constitutive elements:
- The *contents* (information, …), organized following a *material* structure (disposition, page setting,…) which is often enough for a first and quick reading, and a *logical* structure (title, author, date, abstract,…)

which brings some information on the intellectual organisation of the document.
- The *container* (support, medium), which determines the manner in accessing the information.
- The *context of production*, which relates the publication design. This context plays an essential role in the reading process of the document and it can be found as much in the content than in the container.

Contents as well containers, as such, do not enable the easy expressing of the context of production, and therefore the genre (historical or not), of a document. Only published information in the frame of the digital document identifies and determines the context of production and the genre. Sometimes the context of production is difficult to define and so the best way to enlighten it is to refer to a community of practices[12].

With this definition of the digital document, the context of production reveals the genre and shows the quality of the documents dedicated to EHST. We will thus give some criteria:
- The first and major one is the availability of primary sources. If there is no primary source, it becomes difficult to ensure that the document is within the frame of the history of science. The sources have to be contextualised and explained.
- The second one is the use of secondary sources. These documents may help to contextualise and to understand the scientific problem.
- The kind of media: texts, pictures, audio, video, …
- The possibility to make a simulation, to create experiments
- The opening to a new view point and to other historical facts or problems (with hypertexts, links,…).

Then, these criteria can be connected with the aims of EHST digital documents. Is the document about the nature of science? Does it refer to the macro or the micro history? Is it based on the history of a concept or does it link science and society? Does it deal with scientists' biographies or controversies, and so on and so forth?

Guidelines for digital documents in HST for the use in IBST

Now we can propose the main steps of an analysis enabling us to discuss if a resource in a website can be characterized as a historical one and then can be used for the building of an Open Problem Science Teaching from authentic historical examples.

General description of the digital document

The aim of this part is to identify the electronic document through its authors, its destination, and its purpose.

- *Who is the author?:* Is he a professional (a researcher in history of science and technology or in education, an educator, …), an autodidact, an unknown person?
- *Who are the targets?:* has the document been made for EHST specialists, for the researchers or practitioners in education, or simply for a "web-surfer"?
- *What is the aim of the document?:* Is it devoted to research, to scientific culture or for teaching?

Description of the historical resources

In order to work with the document as an historical one, it is extremely important to define what kind of sources are used. We can split the sources into two categories: the primary and the secondary ones. The first category concerns documents in which science is the main and central subject. These are mainly printed texts or manuscripts, sketches, diagrams, representations of curves, technical objects or instruments, …). The second one is mainly a view on science (more or less contemporary) in which the author (historian, philosopher, sociologist) give a discourse with distance and science is not an actor. Mainly, the form of the secondary sources is printed texts.

1. *Primary texts*:

    If a primary historical source is available, the main questions are:

    - Who is the author (or the institution) and do we know the date of writing and publishing?
    - In which language is it written? Is it given in its original language or is it a translation? Is it the first edition or an augmented one? In the case of a re-edition, is it a contemporary edition?

2. *Secondary sources*:

    - Who is the author (or the institution), do we know the date of writing and publishing?
    - Is it an extract or an entire source?
    - If possible, what are its relationships with the original texts? Does it help to understand the original documents?

3. *Elements for the understanding of the historical and epistemological aspects*

    Before using the document, it is useful to understand it as well as possible. It must be given some elements to help the reader to have a precise idea of its historical and epistemological aspects. A contextualization of the document is needed in order to identify the nature of the scientific problem raised and the epistemological questions behind it.

    (a) Historical, philosophical and sociological context of the scientific problem

    In order to study it, anyone can ask questions like:

    - Are there biographical elements about the author?
    - Is there an introduction about the social context or a timeline in which one can see the studied object in history?

- Are there any explanations of the reasons for this study and its importance in the micro-history and/or in the macro-history?
- Are there any social connexions between the people mentioned (or not)?

(b) Nature of the historical scientific problem

In some documents, is it possible to specify the kind of scientific or technological topic? In this case, the right questions can be:

- Is the topic devoted to a specific domain (i.e. mathematics) or is it an interaction between two or more fields?
- What is the relationship with some data, i.e. does the author collect the data himself? What kind of treatment does he use, …?
- Does the document explain or propose a model?
- What is the relationship with the theory? Does it propose a new one? Does it refer explicitly to a specific theory? Does it give arguments against a specific theory?
- Is this document a part of a scientific or a social controversy? If yes, what kind of arguments is used?
- Is the scientific or the technological object a case of interaction between science and society?

(c) Nature of Science/Epistemology

For educators, a better understanding of the social and epistemological dimension constitutes a way to improve IBST:

- Does the document help to understand what science is?
- Are there any considerations about the scientific methods?
- Does it give a specific view about science and society?
- Does it give science a social statute?

Archetypal example

As it has been explained above, the use of historical resources raises lots of questions. In the case of IBST, the documents can be available on-line and this kind of publication generates new questions. In this paragraph, we would like to show that even if there are many things to think about, a web-page can easily give its reader all the useful information needed. For this project, let us take the example of al-Sijzihis text about the perfect compass. An experimental web-page has been designed following the precedent criteria[13].

Here is the main structure:

- Title: Al-Sijzi le compas parfait
- Biographical and bio-bibliographical informations: Life and dead dates, places, and main works. (about 150 words) (fig. 1)

Al-Sijzi et le compas parfait

Fig. 1

- Scientific context: Development of algebra in this period. The resolution of equations and the problem of the continuity of algebraic curves. (about 300 words, 1 picture, 1 graph, and some mathematical contents.

- The main historical subject: The perfect compass and historical comments (about 350 words, 1 schema, 2 quotations with bibliographical references) (fig. 2)

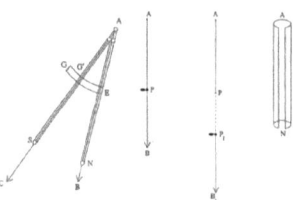

Fig. 2

- IBST around a simulation: with a geometry software, the reader can explore a simulated perfect compass in order to answer some questions about the type of curves that can be drawn by this tool (about 100 words, 2 pictures, 1 simulation, 1 downloadable file) (fig. 3).

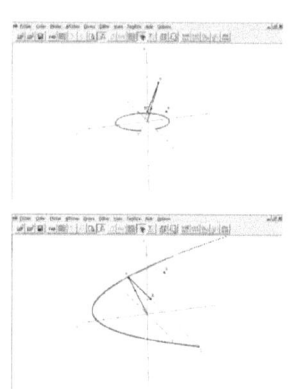

Fig. 3

- Mathematical issues: a historical answer to a problem of continuity, a case of inquiry in mathematics (about 300 words, 1 graph, 1 quotation, some mathematical contents).
- Bibliography: 3 references for the primary texts, 1 article on the subject, 1 secondary source on Arabic mathematics.

Conclusion

Nowadays, on-line publication is quickly increasing quantitatively and it gives many opportunities to create innovative learning sessions. Nonetheless, quantity is not quality and the new technologies such as web 3.0 already point out the risk of losing oneself in an ocean of data. Facing to this situation, historians of science should not stand back. The examples above show that a little vigilance enables us to make a document suitable for IBST with all its historical wealth. The task is not so heavy (3 or 4 paragraphs are often enough) and the community of historians of science should be aware of these questions. IBST is one of the main active topics in didactic research and, in this article, we have tried to show that historians of science have many things to say on this subject. Some new perspectives of collaboration have to be opened, and finally, as the situation does not concern only one country, a European network has to be enhanced in order to share all the experiences.

References

Shepherd, M.; Polanyi, L. (2000) "Genre in digital documents", 33rd Hawaii International Conference on System Sciences, Vol. 3.
http://doi.ieeecomputersociety.org/10.1109/HICSS.2000.926693.

Kanellos, I.; Le Bras, T.; Miras, F.; Suciu, I. (2005) "Le concept de genre comme point de départ pour une modélisation sémantique du document électronique", *Actes du colloque International sur le Document Électronique (CIDE'05)*, Beyrouth, Liban, pp.201-216.

Wenger, E. (1998) Communities of Practice: Learning, Meaning, and Identity, Cambridge University Press.

European Research Programs: http://ec.europa.eu/research/science-society/
Mind the Gap: http://www.uv.uio.no/english/research/projects/mindingthegap/

---

[1] http://ec.europa.eu/research/science-society/index.cfm?fuseaction=public.topic&id=1100
[2] The program ended in 2010, website:
http://www.uv.uio.no/english/research/projects/mindingthegap/index.html
[3] The research group *PaHST* of the university of Brest merged with the Centre F. Viète of the university of Nantes at the begin of 2012: http://www.sciences.univ-nantes.fr/cfv/
[4] Colloque Mathématiques, Sciences exprimentales et d'observation l'école primaire, Table ronde « La démarche d'investigation en mathématiques », Chairman: P. Léna, École Normale Supérieure de Paris, September 28th 2005:
http://www.diffusion.ens.fr/index.php?res=conf&idconf=882
[5] The texts are available in a French translation in *Œuvre mathématique d'al-Sijzi. Volume 1: Géométrie des coniques et théorie des nombres au Xe siècle*, Trad. R.Rashed, Les Cahiers du MIDEO, 3, Peeters, 2004.
[6] http://www.archive.org/details/oeuvresdefermat942ferm
[7] http://math.dartmouth.edu/~euler/docs/originals/E241.pdf
[8] http://www2.lib.udel.edu/subj/hsci/ (last accession 16th November 2009)
[9] http://books.google.com/ and http://gallica.bnf.fr/
[10] Michael Shepherd, Livia Polanyi, Genre in digital documents , 33rd Hawaii International Conference on System Sciences, Volume 3, 2000, pp. 3010,
http://doi.ieeecomputersociety.org/10.1109/HICSS.2000.926693.
[10] See Ioannis Kanellos, Thomas Le Bras, Frédéric Miras, Ioana Suciu, « Le concept de genre comme point de départ pour une modélisation sémantique du document électronique », *Actes du colloque International sur le Document Électronique (CIDE'05)*/ Beyrouth, Liban, avril 2005, pp.201-216
[11] See Ioannis Kanellos, Thomas Le Bras, Frédéric Miras, Ioana Suciu, « Le concept de genre comme point de départ pour une modélisation sémantique du document électronique », *Actes du colloque International sur le Document Électronique (CIDE'05)*/ Beyrouth, Liban, avril 2005, pp.201-216
[12] A community of practices is defined through three aspects: the borders of its application field, its social existence, its language and the documents used and shared by the members of this community. It is also a group with interactions and learning that develop a feeling of belonging and a mutual engagement. See, for instance, Etienne Wenger, *Communities of Practice: Learning, Meaning, and Identity*, Cambridge University Press, 1998 or his website: http://www.ewenger.com/theory/
[13] http://devittori.perso.math.cnrs.fr/sijzi/Compas.htm This page has been successfully used in a short teacher training session in 2008.

Remarks about ethical specificities of presenting online resources in history of biology for inquiry-based science teaching

Hervé Ferrière

Centre de Recherches et de Ressources pour l'Enseignement et la Formation, Institut Universitaire de Formation des Maîtres de Guadeloupe, Université des Antilles-Guyane ;
Centre F. Viète, Université de Bretagne Occidentale, France;
hferriere@iufm.univ-ag.fr

ABSTRACT: Biology historians often study controversies which involve important moral values. The place of those values in the heart of sciences interests the students, especially those who intend to teach sciences later. Teachers can take advantage of this interest in order to enrich the epistemological reflection and the general knowledge of their students. Because studying past controversies allows to understand the construction of science as well as the thought process of scientists. It also enables to understand the way the history of science is written in the historians diverse schools. Because these schools confront exactly on those moral values and about the intellectual approach of researchers. Then, on line resources must clearly specify theses values in order to help students to define rigorously the exact domain of science. In this respect, the study of the different ways to present a scientific controversy enables to avoid confusions that could be made between science and moral domain.

> *"If we had a real education system, there would of course self-defense intellectual."*
> Noam Chomsky (Baillargeon 2006, p.9).

Presentation

The French Nobel Prize in Medicine, Jacques Monod, has declared that "the true Knowledge ignores moral values": he excluded from his investigation what is the ethics of individual or collective "inherently non-objective", whatever the form: moral, religious or political discourses. Moreover, sciences are intended to "attempt to values" – meaning "the contemporary legacies more or less secularised religious or philosophical myths of past centuries" (Blanckaert 1993, p. 14-45). We often hear that "Sciences are amoral". But does it mean that certain scientific data or scientific applications are devoid of any moral implications? Of Course, not!

Yet, those moral implications are the most problematic, when biology teachers try to disseminate an investigative approach of science, using texts or images from books or on line resources of history of science. But, these values can't be ignored: honesty, rigour, tenacity, patience, respect of other people's opinion and methodological materialism (Ferrière 2011, p. 151-154). R.K Merton added to this list the scientific values: disinterestedness, universalism, scepticism and communism (Merton 1973). Regrettably, these values almost never appear in resources for teachers or students. Because they are simply forgotten or implicit. But, students are first often worried by these valuable questions. And, in front of their questioning, teachers give up using these resources (Ferrière 2008).

To illustrate this work, we have chosen a great controversy. We have made a choice for many reasons that we will explain. In fact, we could choose another example: about the age of Earth between Lord William Thomson Kelvin and some other scientists (like Charles Lyell or Charles Darwin), about the signification of the fossils from the explication of Bernard de Palissy, about the rule of spermatozoon in fecundation during the nineteenth and twentieth centuries or, most difficult, about the humans place in animal classification from the Carl von Linneaus' *Systema Naturae*, but, in order to deal with that problematic, we have chosen the controversy based on experiments about spontaneous generation by Louis Pasteur (Fig. 1) and Félix-Archimède Pouchet (Fig. 2) from 1859 till 1864 (in fact, an old controversy which have been discussed between Needham and Spallanzani, or by Bory de Saint-Vincent).

Fig. 1: L. Pasteur   **Fig. 2:** F.-A. Pouchet[1].

Why this choice?

At first, because we can find speeches given by the two protagonists during this famous controversy on the web[2] in the "*Revue des cours scientifiques de la France et de l'étranger*" (Barot 1864, p. 257-270). Other many on line resources are available (in French): the two protagonists' books[3], their portraits, Pasteur's correspondence and manuscripts[4]. We can read "La vie de Pasteur" - Pasteur's life and works[5] -, a short presentation of the controversy for teachers[6] and listen a play[7]. A very simple comparison of both scientists, their laboratories (Fig. 4 and Fig. 5), plans of experiments carried out during the controversy can be found on the web site of the Académie of Aix-Marseille[8]. We can find too some informations about the social and political context. And, last but not least, we can read on the web some of the most important historical and sociological articles about this subject.

On the one hand, this example is very interesting to study together with students in a global research approach. For us, this approach has four important objectives. The first one is understanding the inquiry-based science methodology. We need to learn again to identify all the stages of a discovery; all the different stages of a scientific approach which is sometimes long and difficult (and this example is maybe the best one because, at the end of the controversy, the scientific answer was not definitive and Pouchet will die – in 1872 - without changing his opinion!). The second one is enhancing the student's scientific literacy because they have lost the habit of reading scholarly texts; they prefer to be given a simple protocol to follow. Too often they see themselves as performers, not like real research scientists. They don't always want to find out, with rigor, patience and tenacity, the different lines of thought in a text. The third one is highlighting student's misconceptions about the nature of scientific knowledge because sometimes, they don't know the exact difference between to "know" and to "believe", between "facts", "proofs" or "arguments". And the last objective is also increasing student's interest and motivation for sciences, history and epistemology.

On the other hand, this debate is really well known and also studied in many books and journals. This controversy is very important for another reason: it also can explain and show to teachers and their students how historians write the history of science, scientists and institutions. In fact, this real intellectual fight between those two men in the nineteenth century is used by the two major schools of thought concerning historians and sociologists of sciences in order to defend their point of view about "science construction", K. Popper's refutation (Popper 2007) and process of research. Of course, there are many schools in history of sciences (Braustein 2008, Gonzalez 2010, Lecourt 2001) but those two conflicting lines of thought are in confrontation concerning the importance of the social, political, religious, ideological and finally moral factors (or background) in the resolution of the controversies. Indeed, the *Sociology of Scientific Knowledge* insists on the role of these factors in the history of science. Whereas the other historians of science judge only the arguments and the scientific proofs. There is also a fight between two views of the world and the sciences: a relativist opinion for the first line of thought and a realistic or skeptical view for the second one.

So, studying controversies is a precise choice: at first we want to show how scientists work and how they prove and defend their scientific ideas and methods. Secondly, we want to understand how historians explain the past and the rule of values – scientific (at first) and moral of course – in the resolution of controversies.

The turbulent history of a controversy

Since 1976 onwards David Bloor's book (Bloor 1976), the work of "Edinburgh school" (Barnes 1974), with the research in the University of Bath9 and in the *Ecoles des Mines* in France10, the study of controversies seems to be the "voie royale" - the "best means" - to study history of science. It's probably one of the best means to motivate students because "noise" and "fury" invade the arena of scientists and, in particular, the spans of the Academy of Sciences in Paris, at the time of the publication in England of Charles Darwin's famous work (in 1859). Furthermore, the two protagonists have written their own memoirs and memories (Pouchet 1859, 1860 and 1864, Pasteur 1922, 1951). We can find on the web the Pasteur's correspondence[11]. And, the controversy between Pasteur and Pouchet has been really often studied during the twentieth century and particularly during the last twenty years.

The first report of the controversy has been written in 1907 by G. Pennetier – a Pouchet's supporter and ancient student (Pennetier 1907). And, later, the different studies often summarized the same conclusions advancing the social, personal, political, philosophical, religious and moral aspects of the controversy. This controversy has indeed become one of the best examples of scientific controversy for sociologists of science. They were plenty of scientists to study it and to emphasize the importance of moral values involved in the debate between Pasteur and Pouchet (Dagognet 1967, Farley 1972, Farley & Geisen 1991, Laszlo 1998, Latour 1989[12], Raichvarg 1995, Roll-Hansen 1979[13]). This controversy enabled them to illustrate a central idea of sociology of science: science is a social activity. Other historians and sociologists clearly challenged the conclusions drawn by these authors. They demonstrated that their studies could be read in another way. According to them certain facts have been distorted by some sociologists (Raynaud 1999[14] and 2003) . Also, being a "social activity" is not contradictory with the fact of doing science (Berthelot 2002[15], 2008).

This "controversy in the controversy" is an excellent example for students because they must wonder why this historical episode was so turbulent. They must then find a set of data (historical and scientific) to understand the stakes. Challenges are numerous and most of them concern all the values we were talking about before: rigor, honesty, respect... And then, questions about the existence of God or about the theory of species evolution eventually appear during the study.

The resources we provide to the students must be rich enough (all kind of information) to afford this advanced research. Giving raw texts may not be enough. During this study we can see – paraphrasing Bruno Latour - "science in process", and also, by the same way, the "history of science in process". This controversy between sociologists and historians of science fits to the same scientific rules as during the quarrel of the two scientific from the 19th century. It describes science and history of science being built before our eyes.

But, nevertheless and maybe at first, studying controversies remains a good means to reveal the importance of all the values in scientists work. Then, it is a essential work to exemplify this important problem in the history of biology. Why these two scientists fought so much, during many years, in front of different various sorts of assemblies, in books, newspapers, letters and speeches? To see the triumph of the scientific truth? For the glory or the recognition? To defend a view of the nature, the life and its origin? To defend faiths and philosophic convictions? To defend their own interest and the interest of his scientific and politic community? Maybe for all these reasons.

What are our bets?

Before going on, it is necessary to us to propose our working hypotheses. We think that the history of biology helps the understanding of the scientific inquiry by the students, and, secondly, we are ready to offer some on line resources to them. But, if we only give these resources without explanations, without a rich historical background, it is necessary to make some bets.

The students and the teachers can understand texts, drawings, models or experiments (and sometimes only their narrative) of the past. Then, they can understand from the current theories the most famous ones: theories of evolution and ecology. Finally, they can use them to understand the current challenges posed by globalization of science and the complex ecological and economic organization of the planet. They can understand the complexity of these major theories. They can rebuild the inquiry conducted in their elaboration. But with what sort of resources? They must be able, thanks to online resources, to find the values that guided scientists in the past. This will enable them to understand the values that they will mobilize themselves and defend them in the future during their scientific careers.

Then, after these methodological considerations (related to our teaching philosophy), let's go back to our protagonists. What about their values?

An exemplar controversy for different reasons

The texts of the lecture and debate of those two scientists illustrate some key points of scientific process of investigation. This example helps to show that this kind of investigation is primarily based on the knowledge of the historicity of the concepts of evidence, technical tools and experiments. The text presenting the arguments and demonstrations of the two scientists could be read with an interpretative guide - sometimes formal but effective - showing the different strengths of investigation process.

Pasteur's text also presents an interesting summary of the history of the question from the seventieth century. He quotes four scientists of the eighteenth century: the biologists J.-B. Van Helmont and G. Buffon, the chemists L. J. Gay-Lussac and N. Appert (inventor of a method of food preservation). From his part, Pouchet quotes twenty five authors – a majority of recent (or contemporary) biologists and some physicians or physiologists from Europe - among whom L. Spallanzani, C. Bonnet, J. Needham,

G.R Tréviranus, C. W. Scheele, T. Schwann, J. Ingenhousz, R. Owen, K. E. von Baer, K. Burdach, C. G. Ehrenberg, P. Mantegazza, H. Milne Edwards and the famous physiologist C. Bernard.

There, we see another problem with the history of science teaching. It is also necessary to add a presentation of the historical context and the "characters" of the studied problem. And, we can say that Pouchet mobilizes a lot of biologists' names: a classic argument of authority (not really "honest" and "scientific"). He attacks Pasteur and call him "the Ecole Normale director" (the "simple director" not the "scientist") or the "chemist" (not the "biologist").

Besides the numerous quoted names, there is also a problem with the scientific vocabulary (and rigor). The meaning of some concepts has changed. Others no longer exist today. An overview of the changes of scientific terms is necessary. But, the worst aspect is the domain of philosophical, social, religious and moral values. They have considerably changed. Who thinks today that Pasteur's fought against the spontaneous generation also because he was a religious man? He defended his diverging view of the science and the scientific method (rigor, honesty and patience). But he also fought for his faith. Even if we often present him today in France, as a good example of scientist who advanced step by step, with method, without prejudice nor a priori, like a "non-religious saint", a real "hero of the republican sciences" (after the end of the Second Empire, in 1870, Fig. 3).

Fig. 3: The good "Pasteur" - "Priest"

It is thus necessary to explain the exact nature of the mobilized implicit values, maybe despite both protagonists, in a quarrel of parishes.

But more simply, those texts illustrate the stages of an inquiry-based science learning.

Students will then have to draw up an inventory of the multiple hypotheses tested by the experimental system. Students had to question themselves about fundamental scientific questions, of the past and present, such as about the origin of life or species diversity. At the same time, they must recognize the level of knowledge at this period in order to understand the limits of the demonstrations done by the protagonists.

The posed scientific problem is clear: it's about testing the central hypothesis of the "heterogenists" - the partisans of the existence of spontaneous generation. Then it's about to test the hypothesis concerning microorganisms (animals, plants and "bacterial" – even though this last category of living beings was not clearly understood at Pasteur and Pouchet's time). The following questions arise: how theses living beings appeared? Do they appear in conditions that we can reproduce in a laboratory? Where do they originate from? What are the physical conditions of their existence: do they live in the air, in water or in other living organisms? We try to understand what is their survival condition: at what temperature do they disappear? At the time of this controversy, is it assumed that no living being can survive in boiling broth (today, it is known that some bacteria can survive). We try to find out how they interbreed and how they propagate: do they spread in the air or in water? All testable consequences of these assumptions are clearly stated (despite the fact that the two protagonists had not really and definitively solved the question of spontaneous generation because this controversy will start again in a few years against Pouchet and other scientists like N. Joly, Ch. Musset or A. Trécul). Remarkably, the effects of the controversy also appear out of the domain of science (moral or religious implications).

More broadly, the study of these texts can illustrate the problem of the context of formulation or discovery and the context of justification (Reichenbach 2004, 303-313). It can reveal conceptual backgrounds of scientists work and their thinking. Students can therefore develop historical and epistemological studies - internal and external at the same time - of the two ways of thinking.

For Pouchet, knowledge about microorganisms must participate in a philosophical view of nature. These microorganisms are essential in establishing a global view of origin of life.

For Pasteur, microorganisms are considered in a hygienic and technological vision of the human environment. His point of view is a kind of chemical and mechanistic view of living beings (and of their functioning). It is interesting to note that the philosophical materialism seems to be the Pasteur's fear. So he uses the argument of the rigor, and that he claims that his approach is not based on any ideological assumption. Pasteur is as a "reductionist" and he seeks to disqualify Pouchet (who, in his earlier writings, has attempted to demonstrate that his heterogeneous theory was not antireligious).

Studying directly (or "in live") a scientific demonstration

The various levels of the two texts (demonstrative and rhetoric) are mixed with the specific main points used during a process of investigation (problems, hypothesis, testing hypothesis...)

Many different kinds of arguments are developed to consolidate their respective points of views. Thus, we see the two parties involved in this controversy using arguments and schemes that may appear irrelevant during such an investigation. Find all the arguments is a good exercise, and this sort of exercise already allows students to observe what belongs to science or to ethical values.

It's a surprise for students, we will indeed make use of modern technical process and media. We are going to stage with extras (an assistant and a public witness) and even artificial processes. We will affirm our experimenter status. We have to convince, to "win vote"... Because, new surprise for students, there was a prize awarded by the most prestigious scientific institutions from Paris. This prize was to reward the one who would show the absence of spontaneous generation.

This information has literally fascinated our students when they realized that science was done this way in the past and is still today. They also understood that all these social aspects of science do not prevent science from producing "true and justified knowledge" (Plato's definition of knowledge). We then focused our study on the Pasteur's "stratagems" to invent new scientific practices and gestures a front of public, as well as scientific language and scientific demonstration. Different sorts of arguments can be questioned in this text. We can see the invention of a "scientific language" and a real show of the two great scientists.

So, the columnist who reports the speech of Pasteur gives us technical indications about the demonstration in progress:

> "At this moment, Mr Pasteur makes throw on the board some of these small said spontaneous generations", "Mr Pasteur realizes then this experiment. He takes the flask under the mercury, uncorks it and makes spend the hay in the round-bottomed flask already arranged in advance on the tank with mercury"[16].

Furthermore, Pasteur introduces the assistant who helped him building his demonstration.

There are many data: the phenomena are discussed with clarity and experimental devices are described. The technical instruments and techniques used (round-bottomed flasks, mercurial trough, heating condition, transport mode...) are extremely interesting because they show the experimental methods in a "modern way" of construction and validation of scientific argumentation and conclusion by the scientific community of the time. A round-bottomed flask is a chemist's instrument (and Pasteur is a chemist). It represents the best instruments of the time to isolate living beings. Mercurial trough represents a way to have a "sterilized chamber" or a "decontamination chamber" for Pouchet, not for Pasteur (which prefer to close his flask by doing a "swan-neck").

The narrative of experiment itself is also very rich: we see the connection with environments supposed more healthy (like mountain or campaign) and the construction of a scientific view of nature (ordered, detailed, scrutinized in every details even insensitive part or "inaccessible" directly without instruments or mediator as a microscope). The terms and conditions of the experiments are varied: their presentation is explicit. It is possible to represent them as mathematical models or schematic. And the experimental results are presented in the following statistical form and essentially comparative way.

Fig.4. Pasteur's laboratory[17].   **Fig.5**. Pouchet's laboratory[18].

The observations made "in live" (as in the text of the Pasteur's lecture) are interesting and show the diversity of approaches that can be envisaged. Quotations from other works or references to other past and contemporary scientists can also show the scientific way of work. The modeling by introducing "false" microbes (in fact, it was only dust) at the top of the mercurial trough by Pasteur is an interesting example to work with students. This dust represents Pouchet's error: since, because he introduced microbes in the mercurial when he wanted to close his round-bottomed flasks and to create a sterilized chamber.

The added value of these texts lies in the presentation of such items provided to the public by the speaker. In fact, the text is not a simple transcript of their statement but a little more. It's a real lesson and this fact is very interesting for us. We can maintain their methodological process. And it determines the presentation and approach of the two scientists because it should almost be explicit and emphasized by them. We can also observe the techniques used because there is a whole set of systems implemented for, in the same time, to demonstrate and convince. The "journalist" who tells the lecture given by Pasteur gives many details about the staging of the lesson without affect the scientific rigor of the presented arguments. But telling all the fact, objects and technical tools that Pasteur shows to his public, and not only transcribe his words, gives extra weight to his argument. Thus, we can know also the rhetoric used, the self-staging of the speaker, their pedagogy and their technical methods of presentation (projection, lighting...).

A controversy that enables to talk about tensions between science, ethics and beliefs

The different parts of both speeches can be identified: again, their study shows how to build a process of investigation and how scientists relate it thereafter. The rhetorical devices can then cause a discussion about what kind of communication is employed by the two speakers. We can show it from Pasteur's speech.

The two scientists references, duties and celebrity are explicitly mentioned. This presentation is authoritarian in nature. The scientific caution is on Pouchet's side at the beginning. He was the director of the Museum of Rouen, a well-known discoverer and a good physiologist. He had written popularizing books about fertilization and reproduction of mammals. But, at the end, Pasteur is becoming a great scientist and he has the caution. And modernity too. Pasteur talks from Paris while Pouchet – an older and famous man – lives in province. Pouchet mobilizes his friends and press to defend his hypotheses. But he is not close to the political power. And we should not forget that Pouchet has been Gustave Flaubert's professor. Flaubert who sometimes cruelly mocks professors or scientists of province (Flaubert 1881). This fact stays representative of a part of the public opinion about scientists like Pouchet. Besides, we should remind that the Pasteur's works were effective in medicine and veterinary medicine but only after this controversy: to understand the role of microorganisms in the spread of diseases. His works still stay and become economically and technically profitable (for industrial production of beer or wine). And his works are ideologically and politically useful. They were used to control the environment in the context of industrial revolution and to control the urban population behavior (with hygiene measures).

The role and nature of the audience - which are highlighted at the end of Pasteur's speeches - are also very important for two reasons. First to discuss the construction of the evidence and, second, to show the use of an argument of authority in a scientific presentation. Political context of this demonstration has to be mentioned because the Princess Mathilde Bonaparte (Napoleon III's cousin), but also Victor Duruy (the Minister of Education and a good friend of Pasteur) and the famous writers Georges Sand or Alexandre Dumas (the father) are in the meeting room when Pasteur talks[19]. This is an explicit political support from the government and the "Parisian intelligentsia". This is a religious support too because the second empire is clearly conservative and Catholic. Because Pasteur attacks his opponent about two aspects: scientific aspect but also philosophical, religious and moral aspects (the belief in God). The last sentences of his speech are so clear:

> "Gentlemen[20], great questions pose problems today and sow the seeds of doubt in everybody's mind: unity or multiplicity of the human races[21], origin of Human being since some thousand

years or since some thousand centuries[22]; fixedness of the species or the slow and progressive transformation of the species[23]; Only the eternal matter, outside is the nothingness ; The idea that God is useless[24]: Here are some some of the big questions which provoke the quarrels of the people.

Don't be afraid, I do not claim to resolve one of these serious questions, but next to these mysteries, there is a modest question which is directly or indirectly associated, and I may dare maybe to speak to you about it, because we can do experiments on this question[25], and I studied this subject, I believe it, severely and conscientiously .

It's the problem of named « spontaneous generation ».

Can Matter get organized itself? In other words, organism can be born without parents, without ancestors? That is the question[26].

We have to accept it, the faith in spontaneous generation is very old; [...] I come today to fight against this belief." (Barot 1864, p. 257)[27]

This extract answers to the first sentence which declares that the subject of the controversy is not so important than the "origin of life" or the "proof of the existence of God". Pasteur says in an excess of false humility that his speech is only about a secondary issue: the spontaneous generation. The purely transmissive of two educational devices implemented is also interesting for us: it also shows that both scientists know this kind of pedagogy. It shows also that there is an audience who is emerging for such demonstrations.

The final value of this study is the non-settled debate about the "origin of life" and not only about spontaneous generation. Because it seems that microscopic living creatures don't appear on Earth one day *ex nihilo*. This text is therefore of considerable wealth through a training process of investigation in science. To know the methodology of history of sciences is here irreplaceable. Understanding the process of investigation in science can not be complete without seeing its past implementation by famous scientists such as Louis Pasteur. But, we can not study scientifically those texts without consider protagonists feelings, objectives, faith and values. If not, we miss the complexity and real interest of history of sciences. It doesn't show only method of science, or great scientists characters and their "old but funny theories", it shows how scientists work.

Conclusion

Finally, this example allows students to understand relations between science and moral values. They can observe scientists mobilizing moral values while working but also historians doing the same while writing history. Far from inspiring a form of relativist criticism of science (and especially the humanities such as history), this study clearly identifies the main points of an education based on investigation. Later, students were asked to explain the ideas and guidelines they would keep in their future researches or in teaching: rigor, honesty, respect of others... This analysis was possible and constructive because the proposed resources mobilized explicitly many social, cultural, historical, epistemological and moral aspects. It's necessary to explain and to point them out when we provide both teachers and students with such resources. As for this point is concerned, on line resources are clearly inadequate. However, thanks to their richness and their possibility to link together, it's easy to reconstruct the many contexts in which scientific research takes place. It's so possible to show to students what choices should be made to clearly identify what comes from a scientific work and what comes from moral. As a result, many problems faced by students and teachers could be solved: confusion, relativism, dogmatism or scientism. It would be also possible to reveal all the richness of the inquiry-based science teaching in biology.

References

Baillargeon, N. (2006). *Petit cours d'autodéfense intellectuelle*, Lux, Québec.

Barnes, B. (1974). *Scientific knowledge and sociology theory*, Routledge & Kegan Paul, London.

Barot, O. (1864). *Revue des cours scientifiques de la France et de l'étranger*, Germer Baillière, Libraire éditeur, Paris, 257-270.

Berthelot, J.-M. (2002). 'Pour un programme sociologique non réductionniste en étude des sciences', *Revue Européenne des Sciences Sociales*, XL, 124, 233-252.

Berthelot, J.M. (2008). *L'emprise du vrai, connaissance scientifique et modernité*, PUF, Paris, 63-79.

Blanckaert, C. (1993). *Des sciences contre l'Homme*, Vol.1, Autrement, Paris.

Bloor, D. (1976). *Knowledge and social imagery*, Routledge & Kegan Paul, London.

Braunstein, J.F. (2008). *L'histoire des sciences, méthodes, styles et controverses*, Vrin, Paris, 87-102, 227-241.

Dagognet, F. (1967). *Pasteur sans la légende*, Rééd., 1994, Les empêcheurs de tourner en rond, Paris.

Farley, J. (1972). 'The spontaneous generation controversy (1700-1860): The origin of parasitic worms', *Journal of the History of Biology* 5, 95-125.

Farley, J. & Geison, G. [1974] (1991). 'Le débat entre Pasteur et Pouchet: science, politique et génération spontanée au XIXe siècle en France', In Callon, M. et Latour, B. (eds.), *La science telle qu'elle se fait*, Ed. La Découverte, Paris, 87-145.

Ferrière, H. (2008). 'Réflexion épistémologique sur les rapports entre enseignement et représentation des sciences, formation des enseignants, humanisme et democratie', In *Actes des XIXèmes JIES de Chamonix*, mai 2008, DVD.

Ferrière, H. (2009). *Bory de Saint-Vincent, l'évolution d'un naturaliste voyageur*, Syllepse, Paris.

Ferrière, H. (2011). *L'homme un singe comme les autres, Eléments d'histoire et d'épistémologie pour enseigner l'évolution*, Vuibert-Adapt, Paris, 151-154.

Flaubert, G. (1881). *Bouvard et Pécuchet*, Charpentier, Paris.

Gonzalez, S. (dir.). (2010). *Epistémologie et histoire des sciences*, Vuibert, Paris, 69-91.

Laszlo, P. (1998). 'La possibilité des rencontres obscures', *Revue Alliage*, n°37-38.

Latour, B. (1989). 'Pasteur et Pouchet: hétérogenèse de l'histoire des sciences', In. M. Serres (ed.), *Éléments d'histoire des sciences*, Bordas, Paris, 423-445.

Lecourt, D. (2001). *La philosophie des sciences*, PUF, Paris, 78-90.

Merton, R. K. (1973). 'The normative structure of science, 1942', In N.W. Storer (ed.), *The sociology of science*, The University of Chicago Press, Chicago, 267-278.

Pasteur, L. (1922, *Oeuvres de Pasteur, réunies par Pasteur Vallery-Radot*, tome II: Fermentations et générations dites spontanées. Masson, Paris.

Pasteur, L. (1951). *Correspondance de Pasteur, réunie et annotée par Pasteur Vallery-Radot*, t. II. La seconde étape: fermentations, générations spontanées, maladies de vins, des vers à soie, de la bière (1857-1877), Flammarion, Paris.

Pennetier, G. (1907). 'Un débat scientifique. Pouchet et Pasteur', In *Actes du Muséum d'Histoire naturelle de Rouen*, t. XI, Imprimerie J. Lecerf, Rouen.

Popper, K.: 2007, La logique de la découverte scientifique, Payot, Paris.

Pouchet, F.-A. (1859). Hétérogénie, ou traité de la génération spontanée basé sur de nouvelles expériences. J.-B. Baillière et fils, Paris.

Pouchet, F.-A. (1864). Nouvelles expériences de génération spontanée et la résistance vitale, Paris: J.-B. Baillière et fils, Paris.

Pouchet, F.-A. (1860). 'Genèse des proto-organismes dans l'air calciné et à l'aide de corps putrescibles portés à la température de 150 degrés', *Comptes rendus de l'Académie des Sciences*, 4 juin 1860, t. 1, Paris, 1014-1018.

Raichvarg, D. (1995). *Louis Pasteur, l'aventure des Microbes*, Gallimard, collection découverte, Paris, 36-46.

Raynaud, D. (1999). 'La correspondance de F.-A. Pouchet avec les membres de l'Académie des Sciences: une réévaluation du débat sur la génération spontanée', *European Journal of Sociology*, 40 (2), 257-276.

Raynaud, D. (2003). *Sociologie des controverses scientifiques*, PUF, Paris.

Reichenbach, H. (2004). 'Experiment and prediction', 1938, chap.1, traduction française, In S. Laugier et P. Wagner (eds.), *Philosophie des sciences*, t.1, Vrin, Paris, 303-313.

Roll-Hansen, N. (1979). 'Experimental method and spontaneous generation: The controversy between Pasteur and Pouchet', *Journal of the History of Medicine and Allied Sciences*, 34, 273-292.

Valerry-Radot, R. (1900). *La vie de Pasteur*, Hachette, Paris, 109-127.

---

[1] Two portraits from the web site of the Académie de Marseille (accessed 4 October 2011).

[2] http://gallica.bnf.fr/ark:/12148/bpt6k215070d/f260.image (accessed 4 October 2011).

[3] For Pouchet: http://pds.lib.harvard.edu/pds/view/6649275?n=188 (accessed 4 October 2011) and for Pasteur, see http://www.gallica.fr.

[4] http://www2.biusante.parisdescartes.fr/img/?refphot=anmpx39x0138e&mod=s (accessed 4 October 2011).

[5] http://gallica.bnf.fr/ark:/12148/bpt6k2033085.r=La+vie+de+Pasteur.langFR (accessed 4 October 2011).

[6] http://plates-formes.iufm.fr/ehst/article.php3?id_article=12 (accessed 4 October 2011).

[7] http://www.legrandpublic.fr/spip.php?article768 (accessed 4 October 2011).
[8] http://www.svt.ac-aix-marseille.fr/outils/experimentation/p2.htm (accessed 4 October 2011).
[9] Led by Harry Collins and David Gooding.
[10] Led by Bruno Latour.
[11] http://gallica.bnf.fr/ark:/12148/bpt6k26736r.r=Pasteur.langFR (accessed 4 October 2011).
[12] http://www.bruno-latour.fr/articles/article/038.html (accessed 4 October 2011).
[13] http://jhmas.oxfordjournals.org/content/XXXIV/3/273.extract (accessed 4 October 2011)
[14] http://hal.inria.fr/docs/00/04/91/20/PDF/Pouchet-EJS.pdf (accessed 4 October 2011).
[15] http://www.cess.paris4.sorbonne.fr/dossierpdf/tjmbpnr.pdf (accessed 4 October 2011).
[16] (Barot 1864, p. 258, p. 261). Translation of J.P. Jounet and H. Ferrière.
[17] http://gallica.bnf.fr/ark:/12148/btv1b69274941.r=Pasteur.langFR (accessed 4 October 2011).
[18] Photographies from the web site of Académie de Marseille and from the web site « Gallica » (accessed 4 October 2011).
[19] http://www.tribunes.com/tribune/alliage/37-38/lazlo.htm (accessed 5 October 2011)
[20] We can see that Pasteur seems to talk only to men...
[21] Important question in this period of colonialism and development of the "scientific" study of races. Pouchet is close to partisans of a multiple origin of the humanity – like, for example, J.-B. Bory de Saint-Vincent (1778-1846), who was against slavery but partisan of European colonialism (Ferrière 2009).
[22] We discover men fossil at Neandertal in 1856.
[23] Charles Darwin published his famous book in 1859. All the great problems of the time are quoted by Pasteur. He knows that theiy are all linked to the question of the faith in God.
[24] Pasteur, catholic, is chocked by the materialism increase.
[25] An important criterion of scientific and experimental studies.
[26] Does Pasteur quote Shakespeare?
[27] Translation J.P. Jounet and H. Ferrière.

Tackling Mobile & Pervasive Learning in IBST

Jean-Marie Gilliot[*], Cuong Pham-Nguyen[+], Serge Garlatti[*], Issam Rebai[*] & Sylvain Laubé[#]

[*] *Computer Science Department, TELECOM- Bretagne, France; {jm.gilliot, serge.garlatti, cuong.nguyen, issam.rebai}@telecom-bretagne.eu.*
[+] *Faculty of Information Technology, Ho Chi Min Ville, Vietnam.*
[#] *Centre François Viète (EA 1161), Université de Brest, France; sylvain.laube@univ-brest.fr*

ABSTRACT: In this chapter, we investigate in which extent, social media, mobile and ubiquitous learning may enhance main IBST features, known as i) Authentic and problem-based activities, ii) experimental procedures, iii) self regulated learning sequences, iv) discursive argumentation and communication with peers. We consider the different dimensions of such learning, and propose to embed it in IBST sequences thanks a scenario-based approach. We also consider needed extension of current open services available, showing that semantic web may fulfill those needs. A semantic based prototype is presented to validate the approach.

Introduction

At the European level, the lack of interest by students in science or in the scientific careers has led to a call for research projects in science education[1] and the publication of the Rocard Report (Rocard 2007) about a "renewed" Science Education where recommendations promoted an evolution of teaching methods toward Inquiry Based Science Teaching (IBST). Inquiry Based Science teaching may be defined by to engaging students in: i) Authentic and problem-based learning activities which are ill-defined and have several answers; ii) A certain amount of experimental procedures, experiments and activities involving practical experience of equipment and including searching for information; iii) Self regulated learning sequences where student autonomy is emphasized; iv) Discursive argumentation and communication with peers ("talking science").

History of Science and Technology (HST) may be seen as a fruitful approach of IBST, and has been investigated in a previous European FP7 Project, called Mind The Gap (n° 217 725). Its objective was "to stimulate a more engaging and interesting science teaching based on IBST principles so that more young people in general, and girls in particular, wish to pursue educations and careers in science and technology"[2]. Importance of Technology Enhanced Learning was acknowledged in this project as work package 5[3] was dedicated to the role of ICT in IBST (Gueudet, Bueno-Ravel et al. 2009, 2010, Laubé et al. 2010). Engaging students in the abovementioned learning activities requires defining relevant problems and corresponding scenarios that enable learners to achieve those activities, according to teacher's didactic intentions.

Nowadays, Digital Literacy has to be integrated in the focus. Web 2.0 is the adequate environment to foster digital literacy and to support IBST. Digital Literacy refers to high levels of proficiency of knowledge. DIGEuLit project (Martin 2006) proposes the following definition of Digital Literacy:

> "Digital Literacy is the awareness, attitude and ability of individuals to appropriately use digital tools and facilities to identify, access, manage, integrate, evaluate, analyze and synthesize digital resources, construct new knowledge, create media expressions, and communicate with others, in the context of specific life situations, in order to enable constructive social action; and to reflect upon this process."

Interestingly, Gilster (Gilster 1997) emphasizes critical thinking rather than technical competence as the core skill of Digital Literacy. Critical thinking is rather central in any Inquiry based methodology.

From an educational perspective, social media applications including blogs, wikis, rich media sharing, etc. fit well with socio-constructivist learning approaches as they provide spaces for collaborative knowledge building, self-regulated learning sequences, discursive argumentation, communication with peers and reflective practices. Technology enhanced learning system have to support constructivist approaches for IBST. Learning Management Systems are not flexible enough to support learning as a cognitive and constructive process (Chatti, Jarke et al. 2007) and lack personalization. Mobile/pervasive computing promotes authentic and problem-based learning activities and practical experience because tools, communications and information access are available at any place, at any time by means of mobile devices and internet access. As we show in this chapter, the combination of social media applications (also called Personal Learning Environments) and mobile/pervasive computing can increase these constructive learning processes in HST scenarios.

Our viewpoint is that future pervasive learning environments will be based on contextual adaptation of pedagogical activities and resources to provide on the fly the distributed software environment, composed of appropriate standards tools, to enable the fulfillment of proposed or chosen assessable activities in a social environment. We are more particularly interested in collaborative and pervasive inquiry-based science teaching approaches. Inquiry-Based Science Teaching in the context of History of Science and Technology provides a rich context for pedagogical scenarios (Gilliot and Garlatti 2009)[4].

In this chapter, we aim at proposing how to enhance scenarios as collaborative and pervasive scenario, and how semantic facilities may ease the learning process. We firstly show how pervasive and social learning environment can support inquiry-based science teaching. Secondly, we describe an existing IBST scenario. Then we examine how collaborative and pervasive tools can enhance learning experience, involving mobile and distant collaboration. Finally, we describe our current semantic integrated prototype tool, before concluding.

Pervasive and Social learning environment as support for IBST

In this section, we analyze three technologies enabling us to enhance IBST scenarios and how their integration complements each other. The benefit of each technology is analyzed in depth and we show how their integration enables us to provide a richer learning environment. First of all, we explicit the main features of the social environments viewed as personal learning environments. Secondly, we show how mobile devices, connected to Internet, facilitate information access. Thirdly, context-aware learning is analyzed according to pervasive computing facilities. Finally, the integration of the three technologies is detailed.

Social Environments as Personal Learning Environments

Socialmedia applications are widespread and have already gained acceptance in learning giving raise to the concepts of e-learning 2.0 (Franklin and Harmelen 2007). Personal Learning Environments (PLE) have emerged from the combination social media applications to support learning. Chatti, Jarke et al. (2007) and Siemens (2006) advocate for personalized and social environments. Indeed, social media applications tools offer new opportunities for users, learners and enable new collaborative activities that are suitable for learning. Mohamed A. Chatti defines PLE as follows: "A PLE is characterized by the freeform use of a set of lightweight services and tools that belong to and are controlled by individual learners. A PLE driven approach does not only provide personal spaces, which belong to and are controlled by the user, but also requires a social context by offering means to connect with other personal spaces for effective knowledge sharing and collaborative knowledge creation." In other words, PLE helps gathering tools, communication and social facilities to foster self-regulated learning sequences by student, discursive argumentation and communication with peer. It is the technical support of self and group work organization, should it be defined by the teacher or collaboratively.

Mobile Computing to enhance Information Access

In computer science, mobile computing provides an ever-present "device" that expands our capabilities - by reducing the device size and/or by providing access to computing capacity over the network (Lyytinen and Yoo 2002). At the same time, it provides camera and microphone, which enables capture of User Generated Content (sometimes geo-localized), witnessing current situations, experiments or activities. (Bannan, Peters et al. 2010) propose an example of Inquiry based learning with mobile devices. Mobile learning is not just about learning at anytime, at any place and in any form using lightweight devices, but learning in context and seamless learning across different contexts (Balacheff 2006) (Sharples, Taylor et al. 2006) (Vavoula and Sharples 2008). It is best viewed as mediating tools in the learning process (Sharples, Taylor et al. 2006). Mobile devices are especially well suited to context-aware applications in which learning may occur in location and time, which are significant and relevant for learners. Mobile learning is a social process that links learners to communities, people and situations. Thus, it can increase the PLE potentialities because it will be available at anytime and at any place.

Pervasive Services to enable context aware learning

Generally, mobile learning systems do not have the capability to inquire, detect and explore their environments (context is implicit). In pervasive computing, the computer has the capability to inquire, detect and explore its environment to obtain information and to dynamically build environment models. We consider pervasive and ubiquitous learning as an extension to mobile learning where the environment and context roles are emphasized. Many definitions of pervasive and ubiquitous learning are given in the literature (Bomsdorf 2005), (Hundebol and Helms 2006), (Jones and Jo 2004), (Siobhan 2007). We can cite the following one "Pervasive learning environment is a context (or state) for mediating learning in a physical environment enriched with additional site-specific and situation dependent elements – be it plain data, graphics, information -, knowledge, and learning objects, or, ultimately, audio-visually enhanced virtual layers"(Hundebol and Helms 2006).

In pervasive learning, the physical environment is more directly related to learning goals and activities and the learning system is dynamically adapted to the learning context. The physical environment is directly related to learning goals and activities. In history of science and technology (HST) approach, a pervasive learning environment can provide situated learning activities (difficult before, sometimes impossible) that are particularly relevant in a given situation.

Social, Mobile and Pervasive integrated to provide a Rich Learning Environment

In short, Web2.0, mobile and pervasive environments may be integrated to offer rich learning environments. They can reinforce proper scenario based on authentic problems. The table 1 summarizes the support offered by the scenario, the Web 2.0, the mobile devices and the pervasive computing to the Inquiry Based Science Teaching.

The scenario, giving appropriate situations and instructions, will ensure the consistency of the learning experience. Different Web2.0 services may reinforce the IBST characteristics. Many source management tools exist like Zotero, Diigo, etc. that enable documents or URLs storing and sharing. Knowledge construction is a collaborative process, any tool that enable collaborative writing or sketching will be relevant to annotate experimental process. Blogs, e-portfolios or social networks, are different tools that enable self-publication. They are well fitted to support self & groups reflection, enabling to support development's process conducting to self-regulation and autonomy.

Mobile devices, by allowing data generation and access anytime anywhere, and in situation increase all previous tools effects. Finally, pervasive functionalities may ease information access and future data processing by filtering relevant resources according to the current situation.

Adopting any of these facilities is a matter of convenience, for the teachers and students as well. It may be included as scenario's instruction, or it may be socially discussed among the groups or the class itself. As we explained why and how social media applications (or PLE) and pervasive computing might support HST scenario, we can analyze an original HBST scenario and then how to enhance it.

Scenario		Authentic & PBL activities	Experimental and search for information	Self Regulating learning and autonomy	Discursive argumentation & communication with peers
Scenario		Definition of problem	Appropriate instructions	Appropriate instructions	Appropriate instructions
Web 2.0					
	Source management		Searching information		
	Knowledge construction		Collaborative writing		
	Self & group Reflection			Self-publication	
	Group & social interactions		Interaction in activity	Interaction in activity	Interaction in activity, therefore discussion, also
	Self organization			PLE	
Mobile					
	Data capture		Data for experimentation	Data as learning trace	Data as argument
	Web access		Information access during experiments	Organization information in situation	Communication with peers from different places
Pervasive					
	Contextual information		According to activity, place, people, level of proficiency ...	Organization information in situation	Communication with peers from different places
	Recommendation		According to activity, place, people, level of proficiency ...	Organization information in situation	Communication with peers from different places

Table 1: Enhancing opportunities for IBST

An existing HBST scenario

In this section, we choose an example based on a historical problem of technology - the swinging bridge of Brest over the Penfeld (1861-1944). First of all, the complete problem to solve, dedicated to in-service teachers at primary school, is composed in the three following sub-problems:

- Problem 1: understand the industrial landscape in the area of the bridge (Brest is a shipbuilding arsenal for the Navy)[5].
- Problem 2: understand the historical and technological method of problem solving that led to the construction of the swinging bridge[6].
- Problem 3: understand the rotating mechanism of the swinging[7].

For the sake of simplicity, we focus only on the problem 1 in this paper. A typical scenario for an Inquiry-based learning approach, adapted to the problem 1 and the teachers' curriculum is as follows:

1. Problem analysis in small groups (at school): evolution of industrial landscapes.
2. Activation of prior knowledge through small-group discussion (at school): the group determines the prior knowledge) and the knowledge to acquire (concerning bridges and cranes). Printed readings are available.
3. Elaboration of a common strategy to seek information: for instance, why a bridge, where and how? The group explores the information space. It defines the set of activities, achieved in cooperation or in collaboration.
4. Collaborative work and exploitation: a first work may be to localize current and ancient buildings on a map to define the visit. They have to take pictures and interact with the tutor, to gather a maximum amount of information. At school, they have to upload, organize and publish pictures.

5. Collaborative report writing (social knowledge construction), final problem solving in classroom by exploitation of corpus (gathered information, maps, pictures, etc.).
6. Institutionalization (tutor synthesis, in classroom).

This scenario is based on mobile devices (camera) without communication capability and on availability of a trainer during the visit and on a website to share pictures. The instructions given by the trainer for situated activities, at the stage 4 of the above-mentioned scenario, are: from a walk up the Penfeld from Lift Bridge Recouvrance (meeting place on the parking lot of the Tour Tanguy) and by relying on the gathered historical information before the visit about cranes, bridges and views of the arsenal, you have to: i) Photograph all elements of the current landscape with historical aspects about cranes and bridges; ii) Locate the different elements on a current map; iii) Identify and photograph the actual bridges and cranes linked existing bridges and cranes from previous ones: What "continuities"? What "ruptures"? iv) Store and publish information.

Three notebooks with historical pictures about cranes and bridges[8], the industrial landscape in the arsenal of Brest and maps of the port were printed and distributed to the students at school. Four groups of three in-service teachers were constituted and each group got a digital camera. The teacher trainer went with the groups. He interacted with the students to give advices, guidance and answers to queries all along the walk on the explored site. After the return in the classroom tooled up with PC and Internet connection, each group stored and published on a Google site the pictures taken during the walk.

After the visit, the next session (in classroom) was devoted to the exploitation of the corpus of pictures, guided by the trainer: i) to analyze the continuities and the ruptures observed in the history of the industrial landscape; ii) to understand the historical context and the industrial environment of the swinging bridge in Brest. This is a rich scenario with basic mobile technology supporting learning. Collaborative and pervasive computing could enhance the learning experience on site and at school.

Enhanced scenario

According to trainers' needs, we revisit the existing scenario to propose an enhanced scenario, based on personal, mobile and pervasive learning tools. Combination of PLE's and mobile activities provides additional opportunities, enabling collaboration among different sites. Moreover, pervasive learning environments enable situated learning activities that are particularly relevant in a given context.

Firstly, we describe potential relevant activities according to the original HBST scenario in using social media tools and pervasive computing. Secondly, we propose an enhanced scenario, based on this study.

Potential relevant activities

An analysis of the existing scenario is based on the identification of activities, already existing, or that may be potentially added. Many web sites propose list of tools, but few offers comprehensive lists related to pedagogical purpose (for instance, Center for Learning & Performance Technologies (http://www.c4lpt.co.uk/). Yang and Yuen (2009) published a book of successful experiments, proposing usage examples of wikibooks, social bookmarking, virtual worlds, videos or podcasting. We hope that experiments of pervasive learning in IBST/HST may come off to such result. Such a review may conduct to the following potential activities.

Web environment, by providing social media tools and PLE-based organization enables some relevant activities in IBST situations:

- Shared bibliography, in groups, and among groups.
- Synchronous and asynchronous communications with peers.
- Collaborative writing or media construction.
- Peer assessment is possible through online forms.

- Self-reflection, published on blogs or e-portfolios.

Mobile computing gives some basic, but very powerful facilities:
- Data acquisition: for instance, smartphones provide geo-localization that may be added to pictures that may facilitate map marking.
- All groups may communicate with the teacher, even if not at the same place.

As soon as Internet is available, it is possible to divide student groups into subgroups to increase information retrieval and collaboration at distance. A group can be divided into three subgroups: one at the historical site (port), one at the Navy museum and the third one at the local public records. A subgroup can search for specific information in its location, publish it, exchange with other subgroups and discuss with them according to new needs. Consequently, Internet access in mobility may offer some options for groups' organization:
- Data upload: pictures taken may be directly uploaded to any site. This facility may be exploited in groups, but also between groups, according to course scenario.
- According to information gathered on historical site, in museum or in public records, coordination and communications must be done to enhance and to "synchronize" information seeking and knowledge discovery with other group members.

Pervasive computing can provide two functioning modes the push mode and the pull mode. They are defined as follows:
1. In the push mode, the system is able to recommend suitable entities (resources, activities, tools, persons) depending on the current situation[9] without any human interventions (Bouzeghoub and Do 2009; Bouzeghoub and Do 2010; Bouzeghoub, Garlatti et al. 2010)(Pham-Nguyen, Garlatti et al. 2009; Pham-Nguyen, Garlatti et al. 2009).

2. In the pull mode, the group/individual searches for information, activities, learners or tutors. Thus, the groups "express queries" to express specific needs to obtain the relevant resources according to the current situation.

Considering the original IBST scenario (cooperative activities), we can propose three push mode examples to illustrate the u-IBST, as follows:

- Recommend information from Navy museum and local public records retrieved by other group members or subgroup according to the needed domain concepts identified on the port and/or the current activities
- Recommend and provide information from subgroup visiting the port to other subgroups or group members
- Recommend checking some domain concepts missed by students or subgroups on the port.

The pull mode may be used at different steps of the original scenario. A query filters concepts, resources, activities and persons, for example: write queries on relevant domain concepts like "crane", "bridge", etc. according to the current context (activities and localization), on retrieved information from other group members or subgroups according to activities and/or localization.

According to social media applications, mobile computing with Internet access and pervasive computing, we can propose an enhanced scenario according to the original one.

A scenario proposal

The scenario we propose here is a hypothetical scenario. Its aim is to show some opportunities that can be chosen or not by the teacher. The enhanced scenario is defined as follows:

1. Problem analysis in small groups (at school).

2. Activation of prior knowledge:

    a) A first bibliography of relevant definitions, basic resources on the subject is rapidly established on a social bookmarking tool, that enables sharing in a group, annotations and tagging, for example Diigo. Some keywords, or tags will emerge from this first information seeking;

    b) A list of open questions and knowledge to acquire is build on a group shared document;

    c) Each group's member will write a small post in her e-portfolio to relate her observations on group's and owns methodology. He/she will be free to publish it and to allow comments from other students or not;

3. Elaboration of a common strategy:

    a) An activity list is collaboratively defined to work on open questions and assigned to group's members or sub-groups.;

    b) Opportunity of dividing the group to combine site visit and information seeking in navy museum, local public records, and on the web, is proposed by the tutors;

4. Collaborative work and exploitation

    a) The group can communicate by means of synchronous tools, whether chat or vocal, to suggest some information seeking to other group's members;

    b) Data collected on the web will be collected on social-bookmarking tool. Other sources may be digitized, to enable sharing;

    c) Pictures, videos, sounds may be taken, geo-localized, tagged and uploaded on the web;

d) Based on geo-localisation, mobile devices can suggest to consider some noteworthy landscape's element;

e) The tutor can check keywords, conversations, and possibly tracks followed, that relevant concepts or landmarks are treated, and suggest further exploration. It can be done synchronously (remote communication facilities);

f) Then, a meeting is organized with the whole group to analyse all information collected;

g) Each group's member writes a small post to relate their observations on group's and own methodology, and possible improvements;

5. Collaborative report writing. Any collaborative tool, preferably a wiki, or a word processor like Google docs, will fit to this activity. The report is published online and made available to other groups;

6. Institutionalization (tutor synthesis, in classroom).

   a) As concept coverage, activities lists and deliverables, as well as final reports, are available;

   b) Tutor makes a result synthesis and points out difficulties encountered by reusing available information;

   c) Finally, each student will make a final synthesis of the sequence, underlying strengths, weaknesses, and progress made;

As we may notice, scenario instructions, Web2.0 services, mobility and pervasive facilities are strongly intertwined to attain a potentially rich sequence. This sequence takes into account students level and context. Such sequence makes sense in an inquiry-based teaching.

A Pervasive Personal Learning Environment Prototype

According to the PLE philosophy, a user may choose its own tools. We designed such type of environment called, SMOOPLE for Semantic Massive Open Online Pervasive Learning Environment. In this section, we describe its architecture, as a first prototype architecture that fulfils users' needs and ensure semantic data processing. SMOOPLE is accessed through widgets as its own PLE providing relevant information - resulting from queries to the semantic data. The architecture is depicted in figure 4.

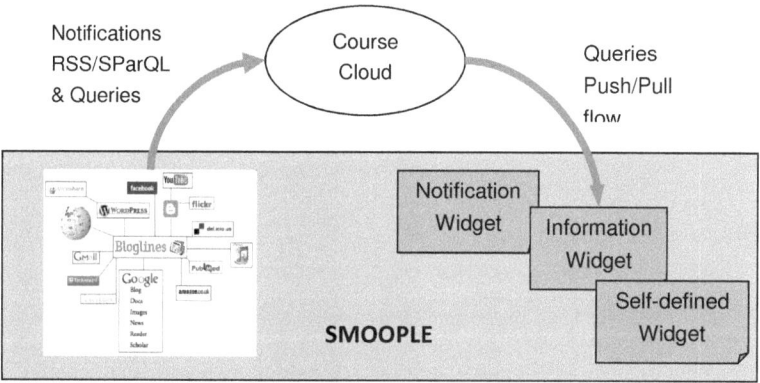

Fig. 4: Abstract Architecture

On the left, we have a set of web 2.0 tools (at present, restricted to Drupal CMS and Wordpress tools). Arrows outline data extraction and making data available to end-user. SMOOPLE manages semantic models, extraction and storing of data produced during activities, making queries available as web services and answering to semantic queries. Semantic Models depict Domain of Interest, User, and Activity descriptions. These functionalities are grouped into a semantic web server based on Jena (Caroll and McBride 2001).

To ensure portability our widgets are implemented using UWA API[10] proposed by Netvibes. At present, we designed different widgets to retrieve information from a learner, a group, the domain model, the tags, the course and, the activities. For activities and groups, it is also possible to create, define and modified groups and activities. The following widgets were used in another curriculum using an IBST approach. Thus, the domain model and the learning objectives are different. The course is about semantic information system. Nevertheless, these different widgets are good paradigmatic examples.

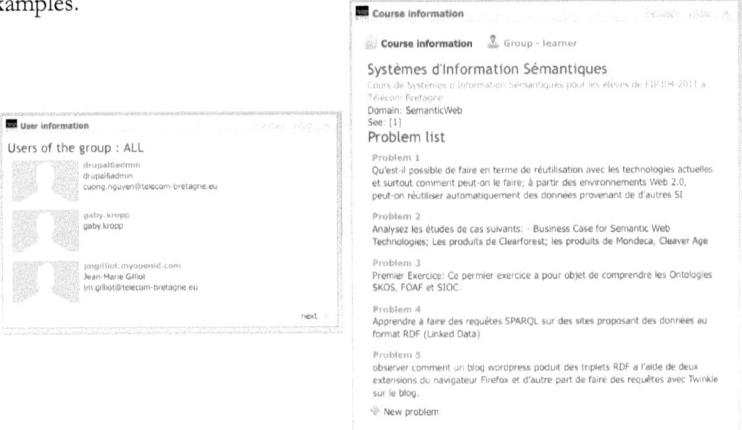

Fig. 5: "User Information" and "Course information" widgets

The widget "user information" provides the list of the registered users on the different social media applications (cf. Figure 5). For a given user, it is possible to get his/her tags and produced content. The widget course information provides the different course elements and the learners groups (cf. Figure 5).

The widget "Tag and domain Ontology", enables us to retrieve content indexed by tags and/or domain concepts (cf. Figure 6). The widget "information for user" provides the content produced by a learner and his/her used tags (cf. Figure 6).

Fig. 6: "Tag Could" and "Information for User" Widgets

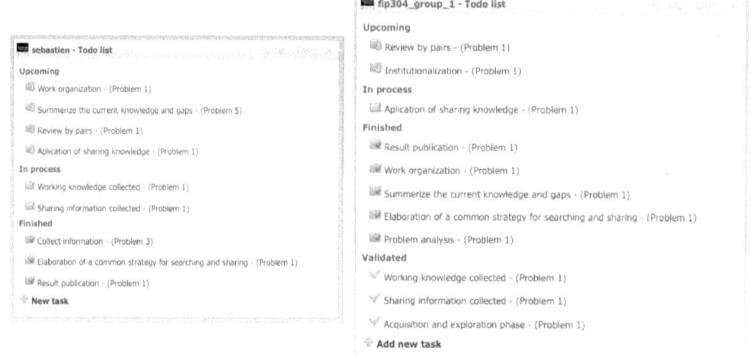

Fig. 7: User "To Do List" and group "To Do List"

The widget "user todo list" enables a user to manage his/her activities (creation, modification of activity state, linking the produced content and the evaluation, etc.) (cf. Figure 7). It is the same for the "group todo list" (cf. Figure 7). We plan to develop widgets for managing peer assessment and self-assessment, recommendations, etc.

Conclusion

Inquiry Based Science Teaching (IBST) renews science education, engaging students in authentic and problem-based learning activities and fostering their autonomy. It is necessary to design new innovative environments coping with IBST fundamental features. Pervasive PLE can provide innovative environments. They can foster collaborative knowledge building, self-regulated learning sequences, discursive argumentation, and communication with peers and reflective practices. Pervasive computing leads to learning at anytime, at any place and in any form, by using small devices, and enables learning in context and seamless learning across different contexts. These capabilities enhance authentic and problem-based learning activities, enabling experiments and practical experience in relevant situations (location, time, etc.). In other words, they can provide rich investigative scenarios incorporating experimental parts (item 1 and 2 of IBST). Reporting services enable students to develop a self-regulated reflective attitude (item 3 of IBST). The use of interactive and collaborative services can encourage activity and arguments between peers (item 4 of IBST).

From a technical standpoint, it is necessary to retrieve relevant information available in distributed tools to be able to reconstruct a complete view of actions, traces, and content of learners, to analyze these actions and to propose relevant recommendations at relevant times and spaces. Our prototype demonstrates that the semantic approach can meet these requirements. Information, available to students, allows them to cast a critical eye on their activities and thus to progress independently (item 3 of IBST). Available to the tutors, it enables them to have relevant information to carry out their tutoring activities.

At present, we developed a first prototype called SMOOPLE as a set of widgets and their corresponding web services to provide different types of information to learners and tutors, in pull mode. We have to go further and to define and implement a context model, an adaptation model and their adaptation strategies to provide recommendations to learners and tutors, in push mode. The next step will be to evaluate this environment in different curriculums.

References

Balacheff, N. (2006). "10 issues to think about the future of research on TEL." Les Cahiers Leibniz, Kaleidoscope Research Report(147).

Bannan, B., E. Peters, et al. (2010). "Mobile, Inquiry-Based Learning and Geological Observation: An Exploratory Study." International Journal of Mobile and Blended Learning , IGI Global2(3):17.

Bomsdorf, B. (2005). Adaptation of Learning Spaces: Supporting Ubiquitous Learning in Higher Distance Education. Mobile Computing and Ambient Intelli-gence: The Challenge of Multimedia, Dagstuhl Seminar Proceedings.

Bouzeghoub, A. and K. Do (2009). Situation-aware adaptive recommendation to assist mobile users in a campus environment. AINA '09: The IEEE 23rd International Conference on Advanced Information Networking and Applications, Bradford, United Kingdom, 2009.

Bouzeghoub, A. and K. Do (2010). Active sharing of contextual learning experi-ences among users in personal learning environments using a peer-to-peer network The 10th IEEE International Conference on Advanced Learning Technologies, ICALT, Sousse, Tunisia (à paraître).

Bouzeghoub, A., S. Garlatti, et al. (2010). Situation-based and Activity-based Learning Strategies for Pervasive Learning Systems at Workplace. Models for Interdisciplinary Mobile Learning: Delivering Information to Students. A. Kitchenham. Hershey PA 17033-1240, USA, IGI Global.

Caroll, J. and B. McBride (2001). The Jena Semantic Web Toolkit. Public api, HP-Labs, Bristol.

Chatti, M. A., M. Jarke, et al. (2007). "The future of e-learning: a shift to knowledge networking and social software " International Journal of Knowledge and Learning3(4/5): 404-420.

Franklin, T. and M. Harmelen. (2007). "Web 2.0 for content for learning and teaching in higher education.", from http://ie-repository.jisc.ac.uk/148/1/web2-content-learning-and-teaching.pdf.

Gilliot, J. M. and S. Garlatti (2009). An adaptive and context-aware architecture for future pervasive learning environments. Workshop "Future Learning Land-scapes: Towards the Convergence of Pervasive and Contextual computing, Global Social

Media and Semantic Web in Technology Enhanced Learning" at ECTEL'09 conference. Nice, France.

Gilster, P., Ed. (1997). Digital Literacy, Wiley.

Gueudet, G., L. Bueno-Ravel, et al. (2009). Technologies, resources, and inquiry-based science teaching. A literature review. Deliverable 5.1, Mind the Gap FP7 project 217725: 1-32.

Gueudet, G., L. Bueno-Ravel, et al. (2010). Guidelines for design on line resources for IBST. Deliverable 5.2, , Mind the Gap FP7 project 217725: 1-32.

Hundebol, J. and N. H. Helms (2006). "Pervasive e-Learning - In Situ Leaning in Changing Contexts."

Jones, V. and J. H. Jo (2004). Ubiquitous Learning Environment: an Adaptive Teaching System using Ubiquitous Technology. ASCILITE, Perth, Australia

Laubé, S. et al (2010) HST, IBST and ICT. Deliverable 5.4, Mind the Gap FP7 project 217725

Lyytinen, K. and Y. Yoo (2002). "Issues and challenges in ubiquitous computing." SPECIAL ISSUE: Issues and challenges in ubiquitous computing45(12).

Martin, A. (2006). "DigEuLit: Concepts and Tools for Digital Literacy Develop-ment" e:LIT 2006 Special Issue5(4).

Pham-Nguyen, C., S. Garlatti, et al. (2009). "An Adaptive and Context-Aware Scenario Model Based on a Web Service Architecture for Pervasive Learning Systems." International Journal of Mobile and Blended Learning (IJMBL)1(3): 41-69.

Pham-Nguyen, C., S. Garlatti, et al. (2009). "Pervasive Learning System Based on a Scenario Model Integrating Web Service Retrieval and Orchestration." International Journal of Interactive Mobile Technologies (iJIM)Vol 3(2): 25-32.

Rocard, M. (2007). Science education now: a renewed pedagogy for the future of Europe European Communities.

Sharples, M., J. Taylor, et al. (2006). "A Theory of Learning for the Mobile Age (preprint)." From http://www.lsri.nottingham.ac.uk/msh/Papers/Theory%20of%20Mobile%20Learning.pdf.

Siemens, G. (2006). "Knowing Knowledge."

Siobhan, T. (2007). Pervasive Scale: A model of pervasive, ubiquitous, and ambient learning. An International Workshop on Pervasive Learning, in conjunction with Pervasive 2007, Toronto, Ontario, Canada.

Vavoula, G.-N. and M. Sharples (2008). Challenges in Evaluating Mobile Learning. mLearn 2008, the international conference on mobile learning, University of Wolverhampton, School of Computing and IT, UK.

Yang, H. H. and S. C. Yuen (2009). "Collective Intelligence and E-Learning 2.0: Implications of Web-Based Communities and Networking." IGI Global snippet.

---

[1] see the FP7 "Science in Society" program at http://cordis.europa.eu/fp7/sis/about-sis_en.html

[2] http://www.uv.uio.no/english/research/projects/mindingthegap/about/index.html

[3] http://www.uv.uio.no/english/research/projects/mindingthegap/Deliverables/index.html

[4] see for example athttp://plates-formes.iufm.fr/ressources-ehst/spip.php?rubrique17

[5] http://plates-formes.iufm.fr/ressources-ehst/spip.php?article17

[6] http://plates-formes.iufm.fr/ressources-ehst/spip.php?article18

[7] http://plates-formes.iufm.fr/ressources-ehst/spip.php?article24

[8] http://plates-formes.iufm.fr/ressources-ehst/spip.php?article17

[9] A situation is a subset of properties accessible from the context at a given moment (localization, device, current activity, etc.)

[10] http://dev.netvibes.com/

Patrimonial Traditions Meet Educational Preoccupations: The Interpretive Shift of the Accessibility Requirement

Ioannis Kanellos

Department of Computer Science, TELECOM Bretagne, France;
ioannis.kanellos@telecom-bretagne.eu

ABSTRACT: This contribution discusses the concept of accessibility to technical, scientific and generally cultural resources from a hermeneutical point of view. It is organized as follows. In the first section we present the underlying technological argument that ensures the convergence between patrimonial and educational logics. In the second, we discuss the very interpretive nature of the accessibility problem; we particularly try to show in which sense accessibility to resources is still an understanding, an interpretation and, finally, a reading problem. In the third section, we show the way this vision meets user-centered design; we especially illustrate proximity between notions of user adaptability, interpretation strategies and reading-centered concerns in accessing collections of resources. We then argue about the role of ontologies in implementing such a conception; in particular, we try to give methodological issues for adequate service building. We finally illustrate our arguments with an example: it is borrowed from a visitor-centered thematic virtual museum concerning Byzantine iconography.

Introduction

It becomes a commonplace of our modernity: by introducing a new technological paradigm, information and communication technologies (ICT) modify our practices and, henceforth, our ways of thinking and acting (Wenger, 1998). Almost all domains are affected; new scientific practices emerge here and there at increasing rates finding grounds on innovative treatments of digital material. In particular, ICT bear fresh expectations as far as accessibility to resources is concerned. In our modern digital lives, ICT seem to bring a bright potential able to reduce the distressing and costly break of social inconsistencies in accessing artistic, scientific or technical resources (Deloche, 2007). We breathe in a post-Gutenberg era where contents are transfigured into pieces of information, shifting towards forms that are more and more emancipated from any notion of support. Indeed, the object, understood under its usual and rather naive materiality, steps aside in favour of its digital representation, which guarantees optimal conditions for information transactions. In today's semiotic economies, materiality is not always relevant; production and reception of contents becomes equivalent to information treatment. In many real life activities (communicating, working, shopping, playing, learning...), "anytime" and "everywhere" demands are formulated, legitimated and realized, thanks to adequate information structures and managements. Traditional frontiers between real and virtual have moved and continue to move; gradually, hybrid realities replace ordinary ones. Rich media reinforce these confusions; the demarcation line between conventional perceptions of what is virtual and what is real, is not always clear: we are currently evolving rather in a "virtually real" than in a "really virtual" world. What is realizable and the way we think the possible are profoundly transformed.

A noticeable consequence of such mutations is the renovation of our relationship to cultural, scientific and technological heritage. Its digital conversion is certainly primordial. But, under the ICT paradigm, accessibility to such an heritage espouses new possibilities and gives rise to new services. Even more: ICT stand nowadays as the fundamental vector of unification of various social demands addressing heritage and knowledge (Cameron & Kenderdine, 2007). Patrimonial traditions meet educational preoccupations. To store and to exhibit is not sufficient: the demand concerns appropriation and understanding (Caune, 2008). Not rarely, entertainment through heritage. Our society is a "forever young" society where emotion and show penetrate all dimensions of the social being.

Nevertheless, the notion of accessibility is polymorphic; it cannot be circumscribed to technical solutions addressing time, space, security, number of users or cost constraints. If constitution of data bases is the digital equivalent to traditional patrimonial and exhibition practices, the accessibility claim is not limited to data base use: it recovers the ways the administered knowledge may be reshaped and adapted to different usages, contexts, needs and objectives; and, of course, to different learning abilities and ambitions. To this accessibility request, cultural, scientific or technological authorities (whether individuals, groups or institutions) try to give answers with tailored services. We thus observe a generalized need of platforms where resources are displayed in ways that carry out different user expectations. We can understand such a state of things as an "interpretive shift" of traditional patrimonial conceptions: resources are not more sacralised objects encapsulating a unique sense, generally reserved to scholars, experts, connoisseurs or specialists; they rather are items organized into collections that stand as the ultimate substance for unending (re)interpretations by different populations (see, for instance, Arasse, 2004).

The need of dedicated thematic spaces containing collections is clear; but what is less clear is the representational model (i.e. the knowledge type, quality, quantity and structure) that may transform them to multifunctional tools, supporting various services; and therefore, satisfying various accessibility criteria.

Precisely, the aim of this contribution is to discuss some basic issues of architectural and functional requirements for variable deepness knowledge representations, able to support alternative reading strategies of a collection of resources.

The accessibility requirement as an adapted reading possibility

We do insist: fundamentally, accessibility is sense accessibility; i.e., comprehension (explicit or not, conscious or not). As humans, we are lifelong prisoners in sense; during this non-negotiable imprisonment, our main living activity is of semiotic type. Even emotions, feelings or dispositions, ideas, beliefs or values, conceptions, thoughts or norms are firmly subordinated to comprehension; perhaps, they are modalities of comprehension.

By formulating the problem of accessibility as a problem of comprehension, we straightaway put the reading process in the centre of accessibility effectiveness: "better accessing resources" becomes "better reading them"; i.e., better understanding and adapting them to different needs. It has been underlined that different persons, at different times and conditions, with different objectives, under different circumstances and even under different situational parameters and with different cognitive skills formulate different reception demands. Any displayed resource has therefore to be able to furnish pieces of information that fit different accessibility criteria. In other words, resources have to be prepared in a way that could be adapted to various user exigencies. Such a user is, in reality, a "reader".

This last is a recurrent concern: many R&D works try nowadays to give solutions for resource adaptability. What is perhaps less observable in reviewing such works is that their way of thinking, modelling and implementing is—indirectly but clearly—*hermeneutical* (Eco, 1992, 1996; Rastier & Bouquer, 2002). Indeed, the hermeneutical approach focus to the reception circumstances than to some supposedly objective information; it is interested in meaning conditions and procedures than in veracity and verisimilitude, in proof and probity, in exactitude and fidelity to facts. The only recognized reality is the meaning reality. Consequently, its central preoccupation concerns the available understanding potential rather than some allegedly robust truthful (re)presentation. The only evidence of all hermeneutical paradigms is the subject and its understanding impulse. However, such a subject is not a restyled pole of irreducible subjectivity; it is an instance of norm stabilization (what user-centered design calls sometimes "a persona"): the subject operates a synthesis over understanding norms. Such norms can still be studied as generalized phenomena.

Such a shift implies a radical conception of the accessibility requirement:
- accessing a resource means to be able to understand it at her/his own level and in her/his particular context;
    - understanding it, invokes aptitude to interpret; and, finally,

- interpreting it, supposes to be able to read it.

Thus, under a hermeneutical paradigm, where the importance is dislocated to meaning and the human procedures bearing, situating, orienting and adapting meaning, the accessibility requirement is transformed to a reading requirement. The considered resource disappears in favour of its possible readings, and subsequently, its possible interpretations (Pearce, 1994). Practically speaking, the objects of a cultural, educational or technological collection are no more seen as objective entities, the same for everyone, but as fluid interpretation underframes. Same in nature, the accessed objects appear at times quite different, insofar as they are embedded into different interpretation contexts.

This last has critical consequences in the way resource associated knowledge is modelled. Indeed, such knowledge has to correspond to possible interpretational contexts about the collection. At least, to identified, normed interpretation categories and procedures.

The discrepancy to emphasize at this point is that it is impossible not to understand at all; but it is also impossible to completely and objectively understand. There is always a tremendous amount of plausible interpretations—and thus of meanings—of everything, and for everyone. We can so say that, anyway and somehow, the user will finally access to a resource. However, we need more than evidence, insofar as understanding evaluation is tightly related to profile, intention and goal prerogatives. Definitely, accessibility evaluation turns into interpretive satisfaction as well as into service effectiveness. Both, flexible interpretive schemata and efficient services thoroughly depend on pieces of knowledge (ontologies) associated to resources. Perhaps, the general disillusion concerning already set up e-education techniques comes from some underestimation of the sustainability of the hermeneutical paradigm they lie on.

Let us see closer the association of ontologies, services and reading qualities in a profile-centered design in order to obtain augmented accessibility to collections.

Profile-centered design

Such a design makes clear reference to the human-centered systems tradition in R&D (Wobbe, 1992, Rosenbrock, 1989); but it goes further than usual pervasive usability concerns inasmuch as the resource optimization is understood as a reading optimization. The fundamental question in scientific, technical, educational or, generally, cultural heritage is not how users can practically accede to some fixed in advance knowledge on a collection of items, but rather how to assist them to realize resourceful reading strategies. The point concerns means that may actualize their interpretive potential rather than solutions that force them to limit their interpretive behavior and adapt it to preformatted and rather partial knowledge.

A profile-centered design aims at restoring the role of the user as the principal—and perhaps the unique—semiotic pole in the accessibility requirement (Rastier, 2001, Eco 1992). Technically, it tries to furnish realistic issues to the shift that goes from collection, conservation and presentation to a profile-dependant semantic assistance; i.e., from storage and exhibition to adapted reading and understanding satisfaction. In such an approach, the problem is not how to define some external and somehow formal profile; but rather "accommodation" of the collection items to different, attested reading practices associated to different profiles.

In a certain sense, the profile-centered design aims at setting up interactive systems with variable genre and deepness knowledge representation; such a representation tries to respect identified, generic reading practices. Undoubtedly, the notion of profile may admit direct and trivial definitions (especially formal), or absolutely complex, unformalized ones. Nevertheless, what prevails is to understand the profile as an "image" of real reading attitudes and intentions. They are, indeed, many and qualitatively very different from each other.

For instance, we distinguish four reading attitudes, depending on the "time budget" the reader decides to allocate to her/his reading process:

- Scanning (whose rate is some seconds)

- Skimming (whose rate is some minutes)
- Cultivation (whose rate is some hours)
- Sapience (whose rate is some days, some months, some years, perhaps a whole life or even several lives).

As far as the reading intentions are concerned, we may differentiate:
- Pleasure and/or entertainment and/or jaunt
- Information and/or inspiration and/or support (in speaking or writing)
- Recollection and/or meditation and/or introspection
- Social contribution or pretext to sociability

We also could consider the forms of reading process in capturing information (linear, radial, spiral, chaotic...) and qualify them with detail. We can again limit our attention to knowledge retention depending on different readings, etc., etc. (Russo et alii, 2008; Hooper-Greenhill, 2000 and 2009; Doering 1999; CIDOC, 2009; Gray, 2004, between others).

In any case, it seems fruitful and thoroughly synthetic to bring out the effort in combination to implied reading strategies. We can thus distinguish three main reading forms that address equivalent reading profiles:
- Discovery and initiation reading
- Study and refinement reading
- Scholar and deepening reading

There are not, of course, precise limits between these categories. They broadly correspond to the well-known categories of naive, experienced and expert (connoisseur) reader. Their segregation is not simply quantitative: what is mainly different, is the reading progress and the intended effect. For instance, the discovery/initiation is concentrated to main aspects and progress from element to element, without a global vision (that is, conceivably, a final result); here, association is prevalent. The study/refinement considers resources under a particular topic underlying the study target; here, comparison and unification are basic tools of meaning accessibility. Finally, the scholar approach goes much further than usual reading categories; it asks for detailed information, eventually engaging special observation techniques and semiotic relationships gleaned through extended corpora of items, that may come out from external resources or cumulated experience. Obviously, any reader may slip from a profile to another at any moment (which implies, change of the reading nature and strategy), depending on intentional alterations, cultural and educational experiences, cognitive abilities, etc.

It seems already clear that, in order to set up adapted accessibility services, the underlying knowledge model (ontologies) has to demonstrate great plasticity; the goal is always to furnish sustained support for a wide range of reading intentions and attitudes, and offer issues for ceaseless shifts from a type of reading to another. This leads to precise policies in conceiving the ontological undercarriage, insofar as they ensure the foundations of the services planed.

Ontologies (again)

Since twenty years, we have (almost) everything said about ontologies. However, we perhaps did not suitably emphasized the relationship between ontologies and service potential of an application. Indeed, in cases like ours, most of the services we are able to build are ontology-dependant. For instance, various forms of similarity and path directions, user traceability and contextual assistance, statistics on semantic information and reconfiguration possibilities of the system, are only some particular implications of ontology nature and structure. If the knowledge concerning the targeted objects, practices or functionalities is poor, the supported services are limited or quite impossible. Thus, there is no "good" or "bad" ontology in absolute terms. There are only adequately or inadequately conceived ontologies (with respect to service expectations and, implicitly, to adaptability abilities of these services).

In terms of content and meaning accessibility, the quality of the developed ontologies is directly responsible for reading flexibility; and therefore, for system adaptability to different profiles. The user accommodation is mainly realized through specific ontological shaping. This last argues for a particular multi-aspects and multi-levels ontology modelling: the knowledge representation is presented through different points of view (local ontologies of the domain) and variable deepness levels (ontology categories and details addressing different profiles), in order to be adapted to various user backgrounds, practices and expectations. From the standpoint of the three reading genres above, it can be resumed as in the following figure (figure 1).

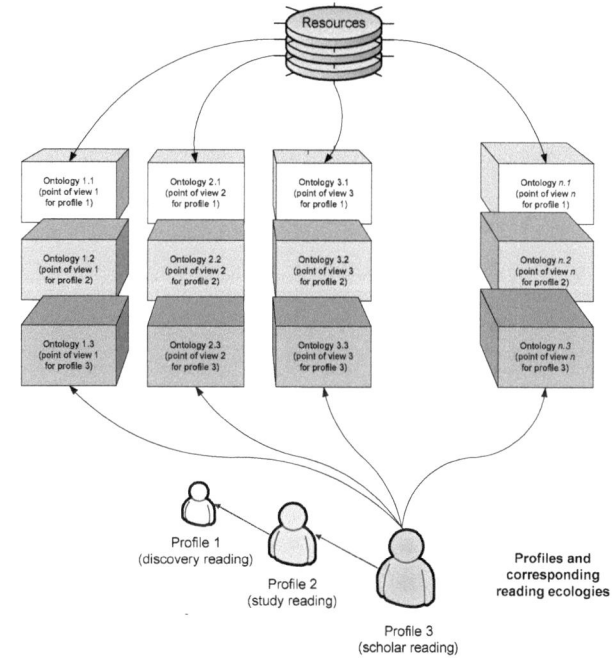

Fig. 1: Resources have to be split into complementary points of view that take the form of local ontologies; different profiles match to different reading practices; they are founded on corresponding various ontology deepness levels.

In this way, knowledge mechanisms and built up services become sensible to different reading formats, various subtleties, and concurrent glances. The above figure is of a generic character: we believe that any accessibility requirement involves a similar knowledge display; at least so long adaptability is concerned as part of the accessibility requirement.

Conceiving services: methodological directions

We give, in this part, an outline of some basic modelling features underlying the interpretive approach of resource accessibility.

- The central concept is the local ontologies that represent the points of view we consider the available resources.
- Such ontologies have to fit the different profiles the designed services address. There must be a specific version of the ontology corresponding to each profile. Thus, an adaptive system has to design and implement $m \times n$ local ontologies, where m is the number of points of view the domain is split into, and n the number of the defined profiles. Clearly, for the main three profiles above, the deepness levels may be conservative reductions of the ontologies of the more advanced profile (scholar profile, for instance).
- Some services are central and seem standard:
    - Be able to trace with accuracy users' paths, when they navigate through resources (in order to furnish to them intelligent and adapted contextual assistance or, perhaps, for governance ends).
    - Edition services (like, for instance, improve (qualitatively or quantitatively) the associated knowledge).
    - Social contribution services (like, for instance, create interest or discussion groups, allow contributions to the ontology schema or content, control global quality or acceptability of technologies or contents, etc.).
    - Serious games and relative participation devices.

The following figure (figure 2) resumes these ideas. It may be seen as a generic representation of what we expect, in a recurrent manner, from an adaptive system offering patrimonial, educational and social functionalities and services.

Implementation issues

In this final part, we sketch the essential implementation choices of a thematic virtual museum we developed trying to take advantage of such ideas (Kanellos, 2009, Kanellos & Daniilia, 2009).

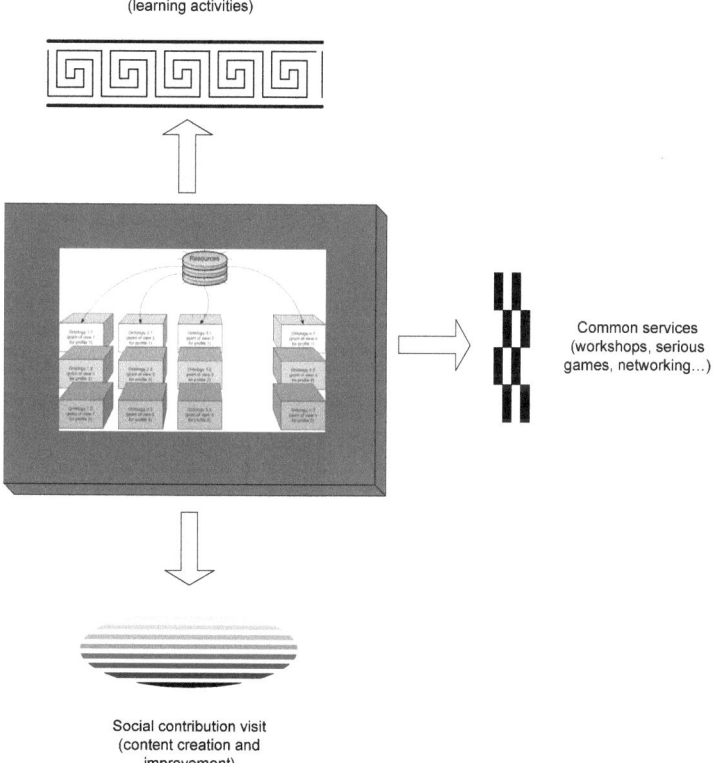

Fig. 2: General architecture of an adaptive platform for unified learning services

In the platform we set up, visits are seen as reading genres; they effectively address different deepness of knowledge for each visiting profile and remain globally interchangeable. This virtual museum, developed with the irreplaceable contribution of the research centre of Ormylia Foundation (www.ormyliafoundation.gr/en), concerns the theme of Annunciation, as it is seen in the Byzantine iconographic tradition (www.annunciation.gr).

The underlying collection of resources contains over 200 high quality digital reproductions of masterpieces (byzantine icons) and hundreds of additional technical images relating to these reproductions. The virtual museum proposes immediately to choose between three reading intentions (cf. above), corresponding to three possible glances on artworks (amateur, student, expert; cf. figure 3); they open to equivalent visiting scenarios.

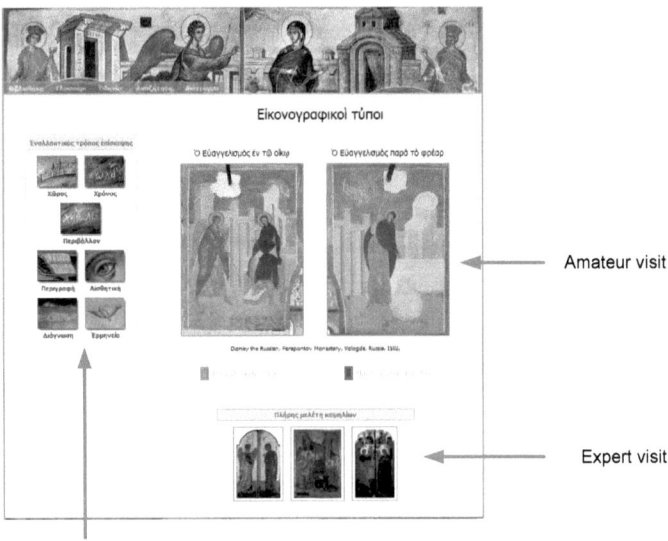

Fig. 3: Regions that correspond to amateur, student and expert readings. Textual material is also associated. Student visit is split into particular study categories; expert visit offers detail investigations of a set of artworks that necessitate particular knowledge background and techniques that enhance the observation conditions.

Amateur profile

The discovery reading (amateur profile) opens to sub-classes of artworks, organized around prototypical icons (i.e. icons with the maximum representational capacity). They are established using formal similarity criteria. Amateur reading progresses, thus, from a prototype to similar icons, presented in an interactive 3D carrousel. The information given at this level of visiting intention is basic (equivalent to the information we generally find in traditional expositions of real museums). However, it is always possible to ask for more information, upgrading the visit protocol, opting for a study or even for an expert one.

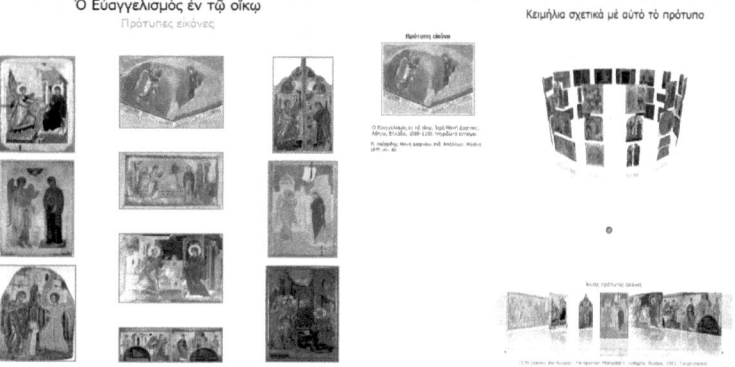

Fig. 4a, 4b: The amateur reading is founded on a gradual discovery path, going from an artwork to another, step by step. The structural principle is broad association between pieces; such an association is not generally explicit. Pleasure and stimulation prevail over synthetic analysis or erudition. In 4a (left) icons represent hints (typical suggestions) for starting an amateur visit. In 4b (right) the interactive carrousel allows to observe pieces one by one with a great comfort (all images are of high resolution). At any moment, the amateur may select some other prototypical icon and modify the visiting plan (flow list at the bottom)

Student profile

The study reading is a class-based operation. The user has an understanding goal and seeks precise information under a synthetic format. For this profile, artworks appear as already organized items. Their "social" nature is more important than their individual specificities: the comparison relationships (i.e., the different ways to put icons with other icons by virtue of an explicative discourse) are here more important. For instance, study reading may ask for classical art history classifications (works of the same period, tradition, painting technique, etc.), combine already categories to construct new ones, or even define original ones, etc. Here, the structural principle of understanding is based on the "interpicturality" relationships (i.e. rationale behind the classification principles of the icons).

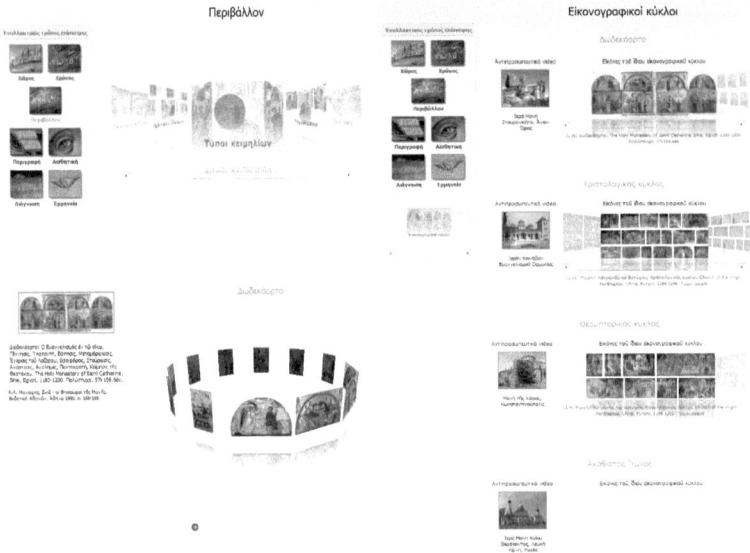

Fig. 5a, 5b and 5c: Some aspects of the study reading. In 5a (up left), we illustrate the subcategories of the context classifications; the choice concerns, in particular, the support in which the selected artwork is realized (wood, wall, tissue, etc.). In 5b (down left), the choice concerns a particular narrative cycle in which the artwork belongs (the 12 great feasts) by means of an interactive carrousel. In 5c (right), we investigate, in particular, such cycles; the left part of the page gives numerous examples of different iconographic cycles and even videos giving expanded information concerning the place and the function of the theme in the iconographic program of a church.

Expert profile

The expert reading goes far beyond traditional categories and insists on details of great technicality. The artwork is here investigated through five points of view:

- Contextualisation,
- Description,
- Aesthetics,
- Physical and chemical analysis and, finally,
- Interpretation.

They correspond to local ontologies that give complementary information of different nature concerning the artwork.

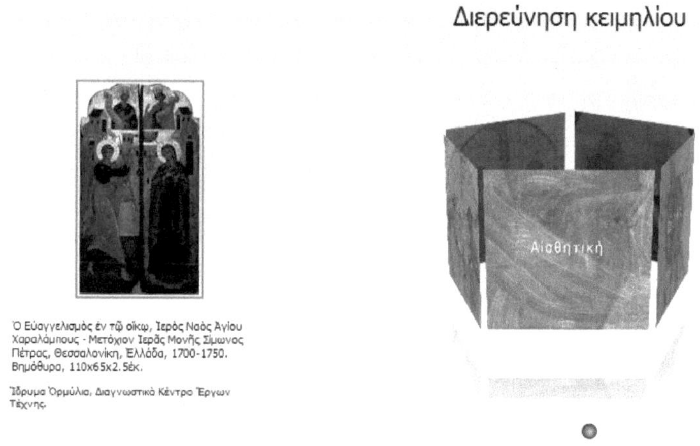

Fig. 6a: The 5 points of view of the artwork are represented by an equinumerous faced interactive carrousel (here stopped to the Aesthetics point of view). The visitor/reader chooses one of the points of view and gleans refined information likely to satisfy her/his plan and sharpen her/his understanding.

The expert reading needs frequently detailed information that may concern either the visible or the invisible building up of the artwork. She/he can formulate demands for restoration, conservation, painting lecture, etc. of any level.

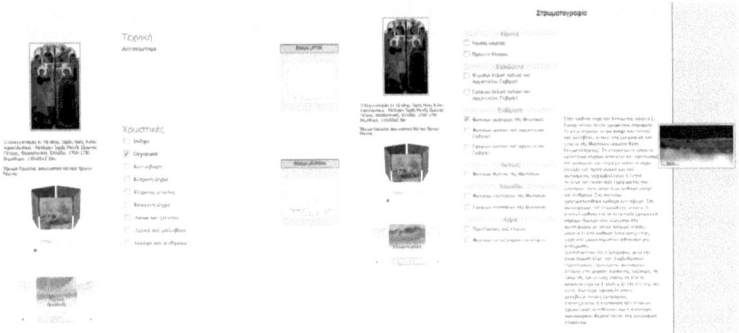

Fig. 6b and 6c: In 6b, we illustrate some aspects concerning the painting technique and the pigments of the artwork; we show the procedure that explicated these last (μRaman technique); we notice that only nine pigments have been used for the whole artwork (here, the demonstration concerns the existence of Orpiment). Spectrographs prove the existence of such pigments. In 6c, we give a stratigraphy example to a given point (vertical section); a technical text describes the obtained image.

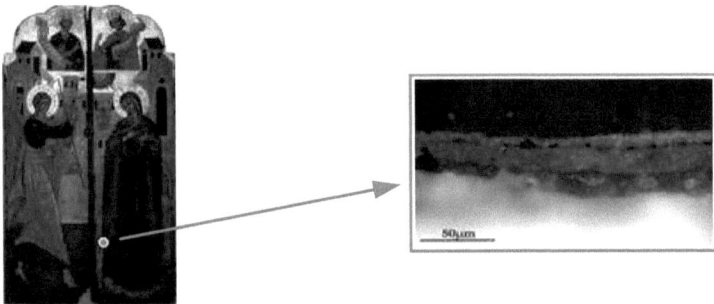

Figure 6d: The point under investigation is localized at the painting surface; the stratigraphy (an image of high resolution as well) allows to precisely study the specificities of the painter, noticeably, the colour layers. The visitor/reader can find here elements that may help her/him to understand the painting mixtures; or issues helping to the identification of the painter and/or the school and/or the period and/or the origin of the artwork, etc.

 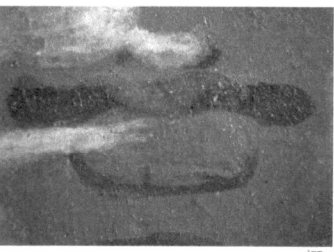

Fig. 7a and 7b: Images of very high resolution extend the observation abilities in the visible spectre (face of the Mother of God and lips of the Archangel Gabriel, respectively).

  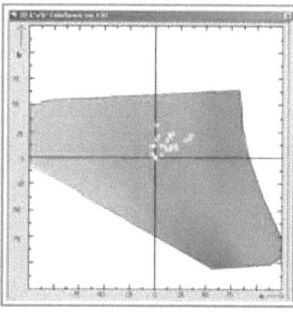

Fig. 7c and 7d: Technical images give also important information concerning invisible parts of the artwork. In 7c we can see an aspect of the first underlying sketch of the artwork (where we find traces of an engraving material, used over a pattern, a usual technique that offers a guide for preliminary drawing lines of the painting). In 7d we have a representation of the visible effect of the painter's palette, obtained from measurements over the indicated points.

Conclusive remarks

Clearly, the patrimonial logic is emblematic insofar as it still remains a main vector of instituting and working social identity. However, standing alone, it is not enough. As expression of a political will, it meets nowadays generalized and crucial educational demands (Caune, 2005). Convergences between educational and patrimonial purposes are nowadays extended and plain; they are motivated by the necessity to offer broad access to important patrimonial values to an as large as possible public. This last concerns artistic and scientific and immaterial and institutional heritage items. Nevertheless, the emergent problems, coming out by unifying patrimonial and educational logics, go further than the replication of respective traditional methods (patrimonial or educational). Nowadays, we assist to a melting pot of practices that tends to set up original spaces (learning spheres or interpretation universes), where knowledge (implicit or explicit) looks for integration issues with entertainment, open social activities or introspection features. Consequently, traditional methods of education have to find innovative ideas to satisfy this renewed demand. We believe that reading and quality of reading are unavoidable in thinking knowledge mutations and accessibility requirements. Technically, this last means that we firmly believe that knowledge adaptability needs refined ontologies, respecting concurrent topics of the available resources. For instance, the ontology of Byzantine iconography upon which our example is founded exceeds 30 000 concepts. Certainly, the cost of digital distribution is almost zero today; but storage, display and interpretation are not unified activities: they all invoke different ways to think accessibility requirements. Definitely, different profiles owe to represent different intentions (and thus, ways) of reading.

Acknowledgments

I would like to express my gratitude to the Ormylia Art Diagnosis Center of the Ormylia Foundation (www.ormyliafoundation.gr/en). Its important know-how on investigation questions and techniques was necessary in setting up this museum. Especially, I wish to thank sister Daniilia, who performed the meticulous technical investigations needed for the expert profile.

References

Annunciation virtual museum: www.annunciation.gr

Arasse, D. (2005). On n'y voit rien. Descriptions. Denoël.

Cameron, F. & Kenderdine, S. (2007). *Theorizing Digital Cultural Heritage*. Cambridge, The MIT Press.

Caune, J. (2005). La politique culturelle initiée par Malraux. *EspacesTemps.net,* Textuel, 13.04.2005. Disponible à: http://espacestemps.net/document1262.html)

CIDOC – ISO standard 21127 (2006). Disponible à: http://cidoc.mediahost.org

Deloche, B. (2007). La nouvelle culture. La mutation des pratiques sociales ordinaires et l'avenir des institutions culturelles. Paris, éditions de L'Harmattan.

Doering, Z. (1999). Strangers, Guests or Clients? Visitor Experience in Museums, *Curator* 42 (2), pp. 74-87.

Eco, U. (1992). *Les limites de l'Interprétation*. Grasset.

Eco, U. (1996). *Interprétation et Surinterprétation*. PUF.

Gray, B. (2004). Informal Learning in an Online Community of Practice. *Journal of Distance Education* 19 (1), pp. 20-35.

Hooper-Greenhill, E. (2000). *Museums and the Interpretation of Visual Culture*. London, Routledge.

Hooper-Greenhill, E. (2009). Museums and Education: Purpose, Pedagogy, Performance. Kindle Edition.

Kanellos, I. & Daniilia, S. (2009). Le concept de musée virtuel thématique: la collection comme visite, la visite comme lecture, la lecture comme stratégie. L'exemple du musée thématique sur l'Annonciation. In *Actes du 12ème colloque international sur le document électronique (CIDE'12)*, Europia Productions, 21-23 octobre 2009, Montréal, Canada, pp. 76-92.

Kanellos, I. (2009). Les musées virtuels et la question de la lecture: pour une muséologie numérique centrée sur le visiteur. *Revue des Interactions Humaines Médiatisées* 10 (2), pp. 3-33.

Pearce, S. (1994). *Interpreting Objects and Collections*. Routledge (Leicester Readers in Museum Studies).

Rastier, F. & Bouquet, S. (éds.) (2002). *Une introduction aux sciences de la culture*. PUF.

Rastier, F. (2001). *Arts et sciences du texte*. PUF.

Rosenbrock, H.H. (1989). *Designing human-centred technology*. Springer-Verlag.

Russo, A. Watkins, J. Kelly, L et Chan, S. (2008). Participatory Communication with Social Media. *Curator*, 51 (1) pp. 21-31.

Wenger, E. (1998). *Communities of Practice: Learning, Meaning, and Identity.* New York, Cambridge University Press.

Wobbe, W. (1992). What are anthropocentric production systems? Why are they a strategic issue for Europe? European Commission, Office for official publications of the European Communities, Luxembourg, rapport EUR – 13968.

PART III

Towards a new strategy for teaching energy based on the history and philosophy of the concept of energy

Manuel Bächtold & Muriel Guedj

ERES LIRDEF University Montpellier 2,
manuel.bachtold@montpellier.iufm.fr, muriel.guedj@montpellier.iufm.fr

ABSTRACT: How can the history and philosophy of science be helpful in teaching the concept of energy? We are currently investigating this topic in a research project that consists of two major themes: the first concentrates on the history and philosophy of the concept of energy, and the second on how to teach energy in classrooms. In this article, we provide an overview of this project and present our reasoning for connecting these two themes. The article then goes on to focus on the teaching of energy, providing a brief summary of the state of play in the field today and discussing current attempts to integrate the history and philosophy of science into teaching programmes. In conclusion, we argue for a larger role to be given to the history and philosophy of science in the teaching of energy and outline the two research areas we intend to develop.

1. Project overview

Today the concept of energy plays a central role not only in the domain of science, but also in the urgent debates taking place on the management of energy and the consequences of its exploitation on the environment. Paradoxically, despite the significance of energy in our society, it remains very difficult to determine what energy precisely is to different observers. Is the energy perceived by the physicist the same as that perceived by the ecologist or the economist? Doesn't each field have its own specific interpretation, resulting in the polysemy of the word? Not only is the concept of energy hard to grasp, but also and correlatively, tricky to teach (see below).

Our research project aims to fulfil two objectives: to clarify the meaning and the role of the concept of energy in the natural and human sciences through a historical and epistemological study, and to rethink the possibilities for teaching energy based on our findings.

The first part of the project will examine the history of the concept of energy in 20th-century physics, as well as its dissemination in other disciplines and in the media, in order to illustrate the diverse understandings of the concept. We have chosen to limit our research to the entrance of energy as a concept in the vocabulary of physics; that is, when the principle of energy conservation was formulated. We will analyse the evolution of the concept of energy in physics with the rise of special and general relativity theories and quantum mechanics, as well as at the level of the properties and operational functions of energy. We will also look carefully at the principle of energy conservation, which gives the concept of energy unifying and predictive power: how has this principle been defined and how has it been used by different scientific communities? We will also examine the way energy concepts have been gradually applied in the experimental practice of physicists. We will analyse the reasons behind why each of the following disciplines, chemistry, biology, geology, psychology, economy and the arts, have introduced the concept of energy (for example, to seek scientific legitimacy or for theoretical needs), the manner in which the concept has been redefined, as well as its uses (literal or metaphorical). Additionally, our investigation will study speeches about energy that have appeared in the media (speeches of science experts, economists, politicians and general public) during decisive historical periods, considering the way the concept has been grasped by the public, and the new meanings it has taken on.

As a parallel study, we will undertake a semantic analysis of historical, philosophical, primary and secondary sources in order to develop a scientific atlas of the multiple meanings and uses of the concept of energy, bringing to light its evolutions, extensions, distortions and inadequate uses and revealing the obstacles to the appropriation of the scientific concept of energy. The atlas will be interdisciplinary and will cover a historical span from the end of the 1800s to the end of the 1900s.

The second objective of this project is to focus on how energy is taught. Because of energy's unifying and operational role, as well as its importance in critical societal discussions in which future citizens will be taking part, teaching energy lies at the core of scientific academic programmes in numerous countries. However, since the 1980s, many science education publications have pointed to a persistent difficulty in teaching the concept of energy. The roots of this difficulty have been well established (substantialism, the abstract aspect of the concept, confusion between forms and transfers of energy, etc.), yet the most appropriate strategy for teaching the concept of energy remains in debate (a mathematical or phenomenological approach, use or rejection of substantialism, etc.) (see below).

This part of the project will concentrate on two aspects: an analysis of energy in education, and the development of a new strategy for teaching energy. The analysis will begin with a historical and comparative study of the evolution and the place of the concept of energy in the school science programmes of several European countries. This will provide the foundation for a critical analysis of the teaching of energy in science education, enabling a rethink of the arguments in favour of concurrent strategies. These analyses will utilise the information collected in the detailed historical and epistemological study carried out in the initial stages of the project.

Based on the findings of these analyses, we aim to design a new strategy that will reform the way teachers are trained. The assumption behind our project is that, in order to teach energy, a teacher must first master its properties and operational functions and be aware of the obstacles to students' assimilation of the concept, and to this end, must have effective and appropriate historical and epistemological training. In this perspective, we aim to develop a European website providing epistemological and teaching resources and to design new strategies for teaching energy in primary and secondary schools.

2. The connection between the history and philosophy of science and science teaching

What is the rationale for a project that investigates the relationship between the history and philosophy of science and science teaching? The fact that both deal with the concept of energy is not a reason in itself. We believe that a detailed study of the historical development of the concept of energy, which analyses all the epistemological implications of this history, is fundamental to conceiving an effective strategy to teach energy.

This assumption is based on the very nature of the concept of energy. Indeed, energy differs from most other scientific concepts because of its richness and complexity at the semantic, epistemological and ontological levels. Semantically, the concept of energy has many meanings depending on the domain of its application (in practical or theoretical fields; in physics, chemistry, biology, psychology etc.). To add further complexity, even within a discipline (e.g. physics), the properties assigned to energy have evolved in correlation with the development of scientific theories.

From an epistemological point of view, the concept of energy encompasses different functions, in particular its unifying and predictive characteristics. Furthermore, the question of the origin of the principle of the conservation of energy is still debated: is it a an *a priori* mathematical structure imposed by the physicists onto the phenomena, is it a purely empirical law, or is it the result of a mutual adjustment between an *a priori* demand of the physicists (looking for an invariant physical quantity) and data of experiences refined thanks to theoretical developments? At the ontological level, the status or scope of the concept of energy also remains disputed: does energy represent a concrete physical reality? If so, how can we describe this reality? How do we interpret the fact that energy can take many forms (kinetic, potential, electromagnetic, etc)? Can we reduce it to a mere mathematical quantity? Although all these questions remain open, science has enabled us to reduce the range of possible answers and to exclude as erroneous some of the common misconceptions (e.g. energy as a kind of fluid).

Because of its wealth and complexity, the concept of energy is very hard to grasp and to teach, and may explain why, as Millar (2005) writes, "The teaching of energy is in a mess". This is why we believe that in-depth knowledge of all its diverse aspects is required in order to reflect on and discuss possible methods for teaching energy. A detailed historical and epistemological study of the concept of energy will be a very useful resource for science teaching.

3. Teaching energy: where we are today and new directions of research

3.1 Energy in school science programmes

Since the 1980s and 1990s, the concept of energy has become very important in the school science programmes of several European countries, which makes it a unique concept to study. The educational context of this period largely explains this fact: the general concern, which dictates new trends in education (and not merely in science education), is that teaching should take account of the major developments in science and in society. In France, for instance, the Bourdieu-Cros report on education (1989) states seven fundamental principles to be applied by the National Council of School Programmes (Conseil National des Programmes), the first of which states: "programmes have to be periodically revised so as to incorporate the knowledge acquired from advances in science and required by societal changes". According to the Bergé report on physics teaching in France (1989), physics is fundamental in this respect:

> *Physics is one of the most important disciplines.* First of all, it has the same strictness in its approach and its reasoning as mathematics. In addition, it is the scientific discipline that provides the main key to understanding the universe in which we live and our everyday experience. Finally, a large ele-

ment of the spectacular progress of our modern world lies, and will lie, on this discipline; for this reason, it is an essential foundation for people [in our] century.

Because of its unifying and predictive power, the concept of energy plays a central role in the very structure of physics. As a consequence, in the last 30 years, it has also become central in school science programmes.

Recently, the Rocard *et al.* report on science education in Europe (2007) points out "an alarming decline" of young people's interest in scientific training (in mathematics and experimental sciences). The report judges the numerous initiatives intended to reverse this tendency as insufficient. The authors emphasise that the reason for this decline in interest lies mainly in the way science is taught and recommend a change in teaching methods: "a reversal of school science-teaching pedagogy from mainly deductive to inquiry-based methods provides the means to increase interest in science." The "deductive method" consists of first teaching the concepts and laws, and then, possibly, but not always, applying them to concrete cases, e.g. through experiments. This method is viewed as too abstract by the authors.

In contrast, the authors indicate a preference for "Inquiry-Based Science Education" (IBSE), which starts with concrete problems to prompt an inquiry, in the process eventually leading to the learning and understanding of scientific knowledge. This inquiry-based learning of science is also intended to provide students with skills to understand social, economic or environmental issues and to take rational decisions:

> Scientific literacy is important for understanding environmental, medical, economic and other issues that confront modern societies, which rely heavily on technological and scientific advances of increasing complexity. However, the key point is equipping every citizen with the skills needed to live and work in the Knowledge Society by giving them the opportunity to develop critical thinking and scientific reasoning that will enable them to make well-informed choices.

In this respect, the concept of energy has a crucial role. It seems essential for young people to acquire an in-depth understanding of this scientific concept in order to make well-reasoned decisions concerning the management of energy and to be able to control the impact of its exploitation on the environment. In recent years, the societal questions involving the concept of energy have become more and more important in school science programmes for this reason. Energy's multidimensional role in today's world requires it to be taught not just as a strictly physical concept, but as an interdisciplinary one.

To this end, our project will closely examine the developments in school science programmes with the aim of developing recommendations on how the concept of energy should be taught and to provide justifications for these recommendations. We will carry out a historical and comparative study in several European countries that will form the basis of our recommendations.

3.2 Energy: a unique subject of study in science education

Since the 1980s, energy has not only acquired an increasing role in school science programmes, but has also become the subject of a large number of science education studies. Researchers in science education largely agree that the traditional method of teaching energy is too dogmatic and abstract (see e.g. Lemeignan and Weil-Barais, 1993). Indeed, teaching energy often amounts to providing students with various mathematical definitions of the concept (e.g. definition of work, kinetic or potential energy), which they have to learn before applying them to concrete cases (this is the "deductive method" referred to in the Rocard report, see above). The origins of the concept of energy and the functions it fulfils are generally not dealt with.

The starting point for science education researchers has been an empirical investigation of students' conceptions of energy and the application (or lack thereof) of the principle of energy conservation (see Solomon, 1982, 1983, 1985, Watts, 1983, Gilbert and Watts, 1983, Duit, 1984, Driver and Warrington, 1985, Agabra, 1985, 1986, Gilbert and Pope, 1986, Trellu and Toussaint, 1986, Trumper, 1990, Trumper, 1993, Ballini et al., 1997, Bruguières et al. 2002). Based on the results of this research, several teaching strategies have been proposed.

In the framework of the very influential constructivist approach, Trumper (1990, 1991, 1993) develops a strategy consisting of eliciting the students' conceptions of energy, and then putting them in situations leading to cognitive dissonance to induce conceptual change.

In the same spirit, Agabra (1986) and Trellu and Toussaint (1986) apply the "objectives-obstacles" strategy (due to Martinand, 1986): first, analyse the students' conceptions of energy so as to identify the different obstacles to the acquisition of the scientific concept of energy, and then devise different specific teaching methods to overcome these obstacles.

Still in the framework of the constructivist approach, Lemeignan and Weil-Barais (1993) and Robardet and Guillaud (1995) propose a teaching method consisting of successive steps that enable the students to reconstruct (with the help of the teacher) the concept of energy, identified as a conserved quantity in a chain of transformations (an "energy chain"). Another and more recent approach based on energy chains has been proposed by Rizaki and Kokkotas (2009).

Some authors, such as Solomon (1985) or Chisholm (1992), believe that the current vocabulary used to talk about energy is responsible for many misunderstandings. Accordingly, they propose to reformulate or simplify this vocabulary.

Yet another strategy deals with the problem of the abstract nature of the concept of energy. As a means to surmount this problem, Duit (1987) and Millar (2005) support a long-term strategy consisting of first introducing energy as a "quasi-material substance" and asking the students to explore the multiple qualitative forms of energy, before coming to a quantitative and mathematical treatment of the concept. Such a strategy is debated, for example, by Warren (1982), who argues that we should avoid reinforcing the erroneous substantialist concept. For his part, Ellse (1988) suggests teaching young students by focusing on "transfers" of energy.

The aim of our project in respect to these various strategies is to make a critical appraisal of the relevance of the arguments and internal coherence of each strategy, to gather information on their feasibility and success at the practical level, and to identify the possible mutual influences between them and school science programmes.

3.3 The history and philosophy of science in the teaching of energy

Although the history of science and, in some respects, the philosophy of science are frequently mentioned in studies about energy in science education, they are often reduced to a limited role and used to support a particular teaching strategy.

For instance, Agabra (1986) refers to the two concurrent models of heat in the 19th century (the substantialist and the mechanistic models) in order to determine their respective operational value, which the author represents using a diagram intended to be an "intellectual tool" to better understand the students' conceptions of energy. Similarly, Bruguière et al. (2002) recounts the history of the constitution of the concept of energy with a view to showing that this concept is in part the product of linguistic connections, and arguing for the development of "conceptual maps" to help teachers when making links between different concepts related to energy, both within a discipline (e.g. physics) and between different disciplines (e.g. physics and geography).

Trellu and Toussaint (1986) use the history of the concept of energy in order to compare two teaching strategies, one based on conservation, the other on the study of transfer processes. They attempt to show that these two strategies, like mechanics and theories of heat in the 19th century, have to be challenged. In other studies, Trellu and Toussaint (1986), De Berg

(1997), and Cotignola et al. (2002) refer to the history of the concept of energy in order to point out difficulties students might face in learning the concept (e.g. confusion about force and energy; confusion about internal energy and heat).

Duit (1987) also makes a parallel between the development of the concept of energy throughout history and its possible development in the student's mind, but with another goal: that of justifying a teaching strategy (see above). He argues that, like the scientists in the 19th century, students may need to first grasp the more "tangible" and understandable substantialist representation of energy, before moving on to consider energy as an abstract concept.

In contrast, Coelho (2009) uses the history of science in teaching science with a quite opposite aim. On the basis of a historical analysis of the works of Mayer and Joule, he supports the idea that energy conservation and transformation can be understood, and hence should be taught, merely in terms of "equivalence" between certain quantities (e.g. heat and work), without appealing to the substantialist conception.

Another use of the history and philosophy of science to support a given teaching strategy is proposed by Rizaki and Kokkotas (2009). According to them, the energy chain strategy (see above) enables us to put forward "the unifying and causal characteristics of energy" identified by a historical study of the concept, and thus to take advantage of the spontaneous causal linear reasoning of the students.

In summary, although all these studies develop distinct teaching strategies, they share the ambition to make use of the history (and to some degree the philosophy) of science as a tool: either to better understand the concept of energy and the possible difficulties students may face in learning it, or to support a given teaching strategy.

3.4 Towards a more ambitious role for the history and philosophy of energy

In our view, the history and philosophy of science should not be reduced to a tool for the teaching of energy. Rather, it should be given a central role and put at the core of the teaching and learning processes. We believe that a historical, philosophical and epistemological perspective provides the opportunity to grasp the constitutive phases of the concept of energy, and thereby, to understand its operative functions, its role in the structure of physics, its domain of validity, and its connections with other scientific concepts. Such an understanding might help to dissipate the perplexity both of learners and teachers when confronted with such a rich and complex concept. The question, then, is how to introduce the history and philosophy of science in teacher training and in student learning?

In our view, teacher training is the fundamental starting point for teaching the concept of energy. To avoid the stumbling blocks (for example, terminology, common preconceptions and mathematical formalism) detailed in the first part of this paper, the first stage of our study will consist of exploring ways to epistemologically prepare the teacher by clarifying the concept of energy and the conservation principle. The objective is for every teacher to grasp the concept in all its complexity and be able to counteract student misunderstandings and comprehension difficulties.

The second stage of our study will consist of conceiving, testing and evaluating teaching methods intended for teacher training. We will also consider how best to evaluate these teaching methods. Finally, we intend to design relevant teaching programmes that can be used with primary and secondary classes.

The aim of the third stage of our study is to develop a European "Energy and Education" website, making use of the new means provided by the semantic web. This website will provide resources on the history, philosophy and teaching of energy, gathered, interrelated and commented by researchers in the three domains. It will be designed for the teachers to grasp the meaning of the concept of energy and the conservation principle, as well as to offer them powerful tools to conceive their own teaching of energy based on the history and the philosophy of the concept.

References

Agabra, J. (1985). Energie et mouvement: représentations à partir de l'étude de jouets mobiles. *Aster*, 1, 95-113.

Agabra, J. (1986). Echanges thermiques. *Aster*, 2, 1-40.

Ballini, R., Robardet, G. and Rolando, J.-M. (1997). L'intuition, obstacle à l'acquisition de concepts scientifiques. Propositions pour l'enseignement du concept d'énergie en Première S. *Aster*, 24, 81-112.

Bergé, P. (1989). Rapport de la mission sur l'enseignement de la physique effectué à la demande de Lionel Jospin. Online at <http://home.nordnet.fr/~dduverney/monsite/niveau3/berge.pdf>.

Bourdieu, P. and Cros, F. (1989). Principes pour une réflexion sur les contenus de l'enseignement. Online at <http://www.sauv.net/bourdgros.htm>.

Bruguières, C., Sivade, A. and Cros, D. (2002). Quelle terminologie adopter pour articuler enseignement disciplinaire et enseignement thématique de l'énergie, en classe de première de série scientifique. *Didaskalia*, 20, 67-100.

Chisholm, D. (1992). Some energetic thoughts. *Physics Education*, 27, 215-220.

Coelho, R. (2009). On the concept of energy: how understanding its history can improve physics teaching. *Science & Education*, 18, 961-983.

Cotignola, M., Bordogna, C., Punte, G., and Cappannini, O. (2002). Difficulties in learning thermodynamic concepts: are they linked to the historical development of this field? *Science & Education*, 11, 279–291.

Doménech, J.-L., Gil-Pérez, D., Gras-Marti, A., Guisasola, J., Martínez-Torregrosa, J., Salinas, J., Trumper, R., Valdés, P. and Vilches, A. (2007). Teaching of energy issues: a debate proposal for a global reorientation. *Science & Education*, 16, 43–64

De Berg, K. (1997). The development of the concept of work: a case where history can inform pedagogy. *Science & Education*, 6, 511-527, 1997.

Driver, R. and Warrington, L. (1985). Student's use of the principle of energy conservation in problem situation. *Physics Education*, 5, 171-175.

Duit, R. (1984). Learning the energy concept in school – empirical results from the Philippines and West Germany. *Physics Education*, 19, 59-66.

Duit, R. (1987). Should energy be introduced as something quasi-material? *International Journal of Science Education*, 9, 139-145.

Ellse, M. (1988). Transferring not transforming energy. *School Science Review*, 69 (248), 427-437.

Gilbert, J. and Pope, M. (1986). Small group discussions about conception in science: a case study. *Research in Science and Technological Education*, 4, 61-76.

Gilbert, J. and Watts, D. (1983). Concepts, misconceptions and alternative conceptions: changing perspectives in science education. *Studies in Science Education*, 10, 61-98.

Lemeignan, G. and Weil-Barais, A. (1993). *Construire des concepts en physique*. Paris: Hachette.

Martinand, J.-L. (1986). *Connaître et transformer la matière*. Bern: Peter Lang.

Millar, D. (2005). Teaching about energy. Department of educational studies, research paper 2005/11. Online at <http://www.york.ac.uk/media/ educationalstudies/documents/research/Paper11Teachingaboutenergy.pdf>

Rizaki, A. and Kokkotas, P. (2009). The use of history and philosophy of science as a core for a socioconstructivist teaching approach of the concept of energy in primary education. *Science & Education*, published online: 5 December 2009.

Rocard, M., Csermely, P., Jorde, D., Lenzen, D., Walberg-Henriksson, H. and Hemmo, V. (2007). Science education now: a renewed pedagogy for the future of Europe. Directorate General for Research, European Commission. Online at <http://ec.europa.eu/research/science-society/document_library/ pdf_06/report-rocard-on-science-education_en.pdf>.

Robardet, G. et Guillaud, J.-G. (1995). Eléments d'épistémologie et de didactique des sciences physiques: de la recherche à la pratique. Grenoble: Publications de l'IUFM de Grenoble.

Solomon, J. (1982). How children learn about energy, or Does the first law come first? *School Science Review*, 63 (224), 415-422.

Solomon, J. (1983). Learning about energy: how pupils think in two domains. *European Journal of Science Education*, 5, 49-59.

Solomon, J. (1985). Teaching the conservation of energy. *Physics Education*, 20, 165-170.

Trellu, J.-L. and Toussaint, J. (1986). La conservation, un grand principe. *Aster*, 2, 43-87.

Trumper, R. (1990). Being constructive: an alternative approach to the teaching of the energy concept, part one. *International Journal of Science Education*, 12, 343-354.

Trumper, R. (1991). Being constructive: an alternative approach to the teaching of the energy concept, part two. *International Journal of Science Education*, 13, 1-10.

Trumper, R. (1993). Children's energy concepts: a cross-age study. *International Journal of Science Education*, 15, 139-148.

Warren, J.W. (1982). The nature of energy. *European Journal of Science Education*, 4 (3), 295-297.

Watts, D. (1983). Some alternative views of energy. *Physics Education*, 18, 213-217.

Historical contexts in mathematics curriculum for secondary school

Iolanda Guevara Casanova

Ins Badalona VII, Badalona 08912 & Institut de Ciències de l'Educació de la Universitat Politècnica de Catalunya, Barcelona 08034; iolanda.guevara@gmail.com

ABSTRACT: The math curriculum in Catalonia includes historical contexts since June 2007. The purpose of this research was to develop some historical contexts of the examples included in the new curriculum in math education. Each item includes a part of history that locates the time, the place and the personages who have studied the subject and another section with activities of learning for the classroom. These are designed from old texts or translations and documented interpretations, to solve them in their historical context and to work to the way of personage that has studied it.

This communication presents a summary of research realised during the 2008-09 academic year, with a study license paid by the Department of Education to perform the work: *The history of mathematics in the new secondary curriculum*.

This license belongs to a set of them that CREAMAT (Resource Centre for Teaching and Learning of Mathematics, Department of Education) promotes to develop resources for the implementation of the new curriculum. At present, these licenses and other materials are in Arc-Cercamat application, designed and built by CREMAT that will open to the public during 2010-11.

References to history in the new curriculum

Using history of mathematics to teach mathematics has been strengthened in the secondary school because there appears explicitly in the curriculum. There is a clear reference to the history in the first of the eleven objectives of the stage and in the introduction of the curriculum:

Understanding mathematics as part of culture, both from the standpoint of history and from the cultural diversity of today's world, and use mathematical competence to analyse all kinds of phenomena of our world and to act as a reflective and critical person in different spheres of life.

At the end of contents each course, there are suggestions for example, of historical approaches to certain contents. At the end of each course content, suggested for example, of historical approaches to certain content. They are intended to show the historical development of mathematics as a science in evolution and subject to change, and also contexts where such contents acquire meaning (DOGC 4915:21928)

Hypotheses and questions developed in the research

The introduction of the historical contexts in math class could be a reality after the publication of the new curriculum but it is still an uncommon practice in our classrooms because the various difficulties involved in the process.

The introduction of the historical contexts in math class could be a reality after the publication of the new curriculum but it is still an uncommon practice in our classrooms by the various difficulties involved similar to those outlined in ICMI Study. (Fauvel & van Maanen, 2000: 8-34).

The research has been designed to find answers to three questions:

- How can we work the historical contexts that appear in the curriculum in the classroom?
- What mathematics do students learn when we introduce historical contexts in the classroom?
- Why historical contexts work helps students to acquire the mathematical competence?

Theoretical framework of research

The theoretical frame of reference for this research is the history, mathematics, and their interrelationships.

On the one hand, the history of mathematics shows the development of mathematical knowledge, explains the processes that have been necessary to reach the current moment (Boyer, 1986), (Joseph, 1996), (Mankiewicz, 2000), (Katz 2007; 2008), (Grattan-Guiness, 2004), (Chemla & Shuchun, 2005), (Grungnetti & Rogers, 2000), (Shuchun & Chemla, 2005) (Grungnetti & Rogers, 2000).

On the other hand, school mathematics are a reflection of what society considers kids must learn, future adult citizens of tomorrow. In the connected global world of today, we accept that mathematics are a tool to understand and analyze reality and at the same time we recognize them as knowledge of high cultural value by themselves, which help to think critically about different realities and actual problems.

How history can contribute into the mathematics? Many mathematicians have been interested in history and in the contribution from history into the mathematics and its teaching. In the last twenty years, some of them have written works that are now obligatory reference in this field, (Katz 2000, 2007), (Barton, 2007), (Barbin, 2000), (Rogers ,2009), (Thomaidis & Tzanakis, 2009), (Boero, 1998).

How can we use history to teach mathematics? Several authors argue that into the use of history to teach / learn mathematics, an important factor to consider is the relationship between the development of students' thinking and the development of history.

They recognize that the introduction of history implicitly serves to the teacher to decide sequences, and how to introduce the concepts; it also provides elements for understanding the production of the students. If added the explicitly history, it can serve the teacher as an element of reference to design activities that include historical situations and it can serve the student as a relative measuring of their difficulties, (Furinghetti & Radford, 2008) (Fauvel & van Maanen, 2000) (Radford & Puig, 2007) (Sfard , 1995).

The reference authors in the theoretical framework and their features

4. Using history in learning

Furinghetti & Radford, 2008:
select topics / time students / consolation
Fauvel & Van Maanen, 2000:
implicit / explicit
Katz, V. J.& Michalowicz, K. D., 2004:
historical modules & activities
Radford & Puig , 2007:
ontogeny / phylogeny
Sfard, 1995:
matt thought / history and philosophy

2. History

Boyer, 1986:
General History
Joseph, 1996:
Other cultures
Mankiewicz, 2000:
Katz, 2007 & 2008:
Grattan-Guiness, 2004:
History and historical legacy
Chemla & Shuchun, 2005:
Chinese mathematics
Grungnetti & Rogers, 2000:
Philosophy, multiculturalism, interdisciplinaritat

1. Teaching/Learning of Mathematics

Niss, 2002:
NCTM, 2000:
Bishop, 1999:
Mason & Wilder, 2006:
Tasks and activities

3. Contributions of the history to mathematics

Katz , 2000 & 2007:
Geometry -> Algebra
Barton , 2007:
Geometry ->Algebra
Barbin, 2000:
History = Solving problems
Occasional use/Learning project
Rogers, 2009:
History = Connexions
Thomaidis & Tzanakis, 2009:
Language, history and mathematics
Boero, 1998:
Realistic mathematics

In this study it has been made explicit use of history, namely that not only served as inspiration and guidance about the sequence of activities but also the history is present in the same activities. We use history to inspire and guide our activities and for sequence elements that help to analyse the development of students' thinking, (Massa, 2003) (Demattè, 2006).

History shows that mathematics have been developed through problem solving, we use history to back the idea that problem solving is the core activities of the classroom (Barbin, 2000).

The results of the investigation

The research presents the most appropriate settings to start introducing the history in mathematics classroom in secondary level. The final form is the production of different materials that can be used independently in the classroom, giving teachers the option to make your own choice, to create the most appropriate for their students and the objectives intended with the introduction of these contexts.

We made a proposal about when to introduce the contexts and how to do it. We explain to the reader the criteria used to decide which subjects have been developed, and we include general information (objectives, how to use them, classroom management, timing, materials and equipment needed). Some of the elements presented are revisions of the items produced by the group of History of ABEAM (Guevara & Massa, 2005) (Guevara, 2008) (Romero, Guevara & Massa, 2007) (Romero, Puig-Pla, Guevara & Massa, 2009).

The components of each element

Each element contains an introduction that discusses some issues relating to the subject. A justification of the choice in relation to the curriculum and the social or cultural relevance of the item. The historical context, the situation in time and space of the introduced personage and his questions. The core of the element, text, problems, demonstration. A reflection about what kind of activities can be generated from that historical context. A specific proposal for activities in the classroom. And finally, the aspects of mathematical competence that are developed, and also the bibliographic references.

In the proposal, we situate the activity in a course and we argue what is the best moment to work on a topic and for what purpose. A part of the elements presented in this project has been implemented in the secondary classroom, in these cases, describes how it was introduced and also included some thoughts on its use.

We enclose the references used for the study of each element because they can be useful for teachers interested in the subject and its implementation in the classroom in order to complete and customize their proposal. They can also be used when some students can work more about the subject.

The elements developed in research

They cover the four courses in Secondary Education, using the examples of the official curricula as a starting point, although it seems reasonable to assume that each context can move up or down a course. They belong to four of the five blocks of curricular content: numerals and calculation, change and relationships, measure, space and shape.

They comprise various moments in history and from diverse cultures. Generally we know very little about other cultures outside Europe and the inclusion of the history from these countries is relatively recent when talking about the history of mathematics. In this sense, we have encompassed mathematics from ancient Chinese and the Indians, besides Egypt, Ancient Greece, the Arab World and Renaissance Europe.

Historical context	curricular block	developed aspects of mathematical competence
1st course		
Numbers and Chinese Counting Board (ca. 250)	Numerals and calculation	Understanding the numbers and the different forms of representation.
Finding π with Archimedes method (ca. 287 a C – 212 a C)	Measure	Applying techniques, tools and formulas to obtain appropriate measures and make reasonable estimations
2nd course		
Euclid's *Elements* (300 aC) and Pythagorean theorem.	Space and shape	Using visualization, mathematical reasoning and geometric modeling in solving problems.
Gou gu procedure in China, Liu Hui (ca. 250).	Space and shape	Using visualization, mathematical reasoning and geometric modeling in solving problems.
Negative numbers in Al-Samaw'al and Italian Abacist (s.XII)	Numerals and calculation	Understanding the numbers and the different forms of representation. Understanding the meaning of operations
3rd course		
The *fangcheng*, Systems of Linear Equations, Liu Hui (ca. 250).	Change and relationships	Representing and analyse mathematical situations and structures using algebraic symbols.
Solving 2n degree equations, al-Khwârizmî (c.780-850).	Change and relationships	Representing and analyse mathematical situations and structures using algebraic symbols.
	Space and shape	Analysing characteristics and properties of geometric shapes and develop geometric reasoning about geometric relationships.
Chinese problems (ca. 250) & Indian ones (ca. 625) using negative numbers.	Numerals and calculation	Understanding the numbers and the different forms of representation. Understanding the meaning of operations
	Change and relationships	Using mathematical models to represent and understand quantitative relationships.

4th course		
Menelaus *Spherical* (ca. 100). Construction with GeoGebra.	Space and shape	Using visualization, mathematical reasoning and geometric modeling in solving problems.
	Measure	Applying techniques, tools and formulas to obtain appropriate measures and make reasonable estimations
Regiomontanus (1436-1476) and the triangles resolution.	Space and shape	Using visualization, mathematical reasoning and geometric modeling in solving problems.
	Measure	Applying techniques, tools and formulas to obtain appropriate measures and make reasonable estimations
The syncopated algebra of Diophantus of Alexandria (ca. 250).	Numerals and calculation	Understanding the numbers and the different forms of representation. Understanding the meaning of operations
	Change and relationships	Representing and analyse mathematical situations and structures using algebraic symbols.

Concluding remarks

If you are interested in using history to teach mathematics you will find that this license contains ideas and situations that exemplify this subject. Ideas to justify this subject and working situations for your students.

All contexts have been successfully used in class to teach maths in a secondary school.

The theoretical framework and justification of history as a tool for teaching mathematics have been recently used in initial training of mathematics teachers.

You can read the license at:
    http://phobos.xtec.es/sgfprp/cerca.php
You can download materials at:

http://apliense.xtec.cat/arc/elements_didactics?body=&title=&field_autoria_value=Iolanda+Guevara

References

ARC-CercaMat: <http://phobos.xtec.cat/creamat/cercamat/> (30/09/2009)

Barbin, E. (2000), 'Integrating history: research perspectives'. In: Fauvel, John & Van Maanen, Jan (eds), *History in Mathematics Education, The ICMI Study*, Kluwer Academic Publishers, Dordrecht/Boston/London, pp. 63-90.

Bishop, A. J. (1999), Enculturación matemática. La educación des de una perspectiva cultural, Paidos (col. Temas de educación), Barcelona.

Boero, P. (1998), 'Teaching and Learning Geometry in Contexts'. In: Mammana, C. & Villani, V. (eds), *Perspectives on the teaching of geometry for the 21st century*, Kluwer Academic Publishers, Dordrecht, vol. 1, 52-61.

Boyer, C. B. (1986), *Historia de las Matemáticas*, Alianza Editorial, Madrid.

Chemla, K. & Shuchun, G. (eds): 2005, Les Neuf Chapitres, le classique mathématique de la Chine ancienne et ses commentaires, Dunod (bilingual critical edition), Paris.

Demattè, A. (2006), *Fare matematica con i documenti storici. Una raccoltaper la scuola scondaria de primo e secondo grado*, a-volume for students, b- volume for teachers, Editore Provincia Autonoma di Trento – IPRASE del Trentino, Trento.

Department of Education of the Generalitat de Catalunya: 2007, 'Secondary Curricula Decree 43/2007'. In: *DOGC (Diari Oficial de la Generalitat de Catalunya)* 4915 – 29.6.2007: 21928) <http://www.gencat.cat/eadop/imatges/4915/07176092.pdf >(11/08/2010)

Fauvel, J. & Van Maanen, J. (eds) (2000). *History in Mathematics Education. The ICMI Study*, Kluwer Academic Publishers, Dordrecht/Boston/London.

Furinghetti, F. & Radford, L. (2008), 'Contrasts and oblique connections between historical conceptual developments and classroom learning in mathematics'. In: English, L. D., & others (eds), *Hanbook of International research in mathematics education*, Taylor & Francis, New York, pp. 626 – 655.

Grattan-Guiness, I. (2004), *History of the Mathematical Sciences*, Hindustan Book Agency, India.

Guevara, I. & Massa, M. R. (2005), 'Mètodes algebraics a l'obra de Regiomontanus (1436-1476)', *Biaix*, 25, 27-34.

Guevara, I. (2008), 'The Menelaus Theorem, The Ptolemy proof (s. I) and the Geogebra construction (s XXI)'. In Hunger, H. (ed.), *Proceedings of the 3rd International Conference of the European society for the History of Science*, ESHS, Viena.

Guevara, I. (2009), *La història de les matemàtiques dins dels nous currículum de secundària*, <http://phobos.xtec.es/sgfprp/resum.php?codi=1864> (30/10/2009)

Grungnetti, L.& Rogers, L. (2000), 'Philosophical, multicultural and interdisciplinary issues'. In: Fauvel, J. & Van Maanen, J. (eds), *History in Mathematics Education. The ICMI Study*, Kluwer Academic Publishers, Dordrecht/Boston/London , pp. 39 - 62.

Joseph, G. G. (1996), La cresta del pavo real. Las matemáticas y sus raíces no europeas, Pirámide, Madrid.

Katz, V.J. (ed.) (2000), *Using History to Teach Mathematics. An International Perspective*, The Mathematical Association of America, Washington.

Katz, V. J. & Michalowicz, K. D. (eds) (2004), *Historical Modules for teaching and Learning of Mathematics*, The Mathematical Association of America, Washington,

Katz, V. J. & Barton, B. (2007), 'Stages in the history of algebra with implications for teaching', *Educational Studies in Mathematics*, 66: 185 –201.

Katz, V.J. (2008), *A History of Mathematics. An Introduction*, Addison Wesley Logman Inc. Reading (3a ed.), Massachusetts.

Mankiewicz, R. (2000), *Historia de las Matemáticas*, Paidos, Barcelona.

Mason, J. & Johnston-Wilder, S. (2006), *Designing and using Mathematical Tasks*, Tarquin Publications & Open University, St.Albans/Milton Keynes.

Massa, M. R. (2003), 'Aportacions de la història de la matemàtica a l'ensenyament de la matemàtica', *Biaix*, 21, 4-9.

Niss, M. (2002), 'Mathematical Competencies and the Learning of Mathematics: The Danish KOM Project. Denmark. In:
<http://www7.nationalacademies.org/mseb/Mathematical_Competencies_and_the_Learning_of_Mathematics.pdf > (09/01/2009).

NCTM (National Council of Teachers of Mathematics) (2000), *Principios y Estándares para la Educación Matemática*, Sociedad Andaluza de Educación Matemática Thales. Proyecto Sur Industrias Gráficas, Granada.

Radford, L. & Puig, L. (2007), 'Syntax and meaning as sensuous, visual, historical forms of algebraic thinking'. *Educational Studies in Mathematics*, 66, 145-164.

Rogers, L. (2009), 'History, heritage, and the UK mathematics classroom', In: Working group 15. *The role of the history of mathematics in Mathematics Education: Theory and Research*, 119-128.
<http://educmath.inrp.fr/Educmath/recherches/actes-en-ligne/1wg15.pdf (08/05/2009)>

Romero, F., Guevara, I. & Massa, M. R. (2007), 'Els Elements d'Euclides. Idees trigonomètriques a l'aula', In: Grapí, P. & Massa M. R. (ed.), *Actes de la II Jornada sobre la Història de la Ciència i l'Ensenyament*, SCHCT, Barcelona, pp.113-119.

Romero, F., Puig-Pla, C., Guevara, I. & Massa, M. R. (2009), 'La trigonometria en els inicis de la matemàtica xinesa. Algunes idees per a treballar a l'aula', *Actes d'Història de la Ciència i de la Tècnica*, Vol. 2(1), 419-426.

Sfard, A. (1995), 'The development of algebra: Confronting historical and psychological perspectives', *Journal of Mathematical Behavior*, 14, 15-39.

Thomaidis, Y. & Tzanakis, C. (2009), 'The implementation of the history of mathematics in the new curriculum and textbooks in Greek secondary education'. In: Working group 15. *The role of the history of mathematics in Mathematics Education: Theory and Research*, 139 –151:
<http://educmath.inrp.fr/Educmath/recherches/actes-en-ligne/1wg15.pdf (08/05/2009)>

Some peculiarities of mathematics teaching in Lithuanian basic school: computer, project, excursions, mathematics history

Nijolė Cibulskaitė

Lithuanian University of educational Sciences, Lithuania

ABSTRACT: The reform of Lithuanian mathematical education, which began at the end of the twentieth century, raised educational goals to teach students to investigate, think critically, solve problems, to apply knowledge in practice. New requirements have brought innovations in the teaching mathematics. That is why teachers have started to develop modern methodology in teaching and learning mathematics. The aim of this article is to present the peculiarities of teaching mathematics to the basic (lower secondary) school's students (age 12 – 17) in the last five years. While carrying out the first two researches in 2004 and 2006, 912 and 1177 students of the V-X forms of the basic schools from different Lithuanian regions were interviewed. The most important conclusion of these investigations was that, though the teachers of mathematics apply modern activities in their practical work, they much more frequently use traditional methods which corespond to the requirements of the teaching the subject. This conclusion signify the insufficient of realisation of new didactical demands in the practice of the basic school. Third research was carried out in 2008. The aim of this study was to establish how often mathematics teachers apply such modern teaching activities as the use of computer, projects making and an excursions arranging in the basic school. It has also been interest in how often teachers use the elements of mathematics history in the lessons. The research involved 882 students of the V-X forms of the basic schools. The research reveals that over the last years mathematics teachers much more frequently apply computer technologies as a modern teaching and learning activity, but they rarely arrange mathematical projects and mathematical excursions; the teachers also rarely use the elements of the mathematics history.

Introduction

Lithuania has been participating in TIMSS (Trends in International Mathematics and Science Study) research and the results of last mathematics achievement testing of the VIII form student's (*Tarptautinis matematikos ir gamtos...*, *2008*) showed: Lithuania is in the 10th position among 49 countries; the results exceed the TIMSS scale average by 6 points. TIMSS payed great importance to putting knowledge into practice, but not to knowledge accumulation. However, the results according to the areas of capacity show that Lithuanian students of the VIII form have demonstrated mathematical knowledge much more successfully (8th place) than using it (9th place) and the worst results - in the field of mathematical thinking (16th place).

The Lithuanian Ministry of Education made national student achievement tests which have shown that the outcome of students' mathematics learning highly dependes on teachers' established methods, the work organization in class, the ability to use teaching tools efficiently (*Nacionalinis VI ir X ...*, *2004; Nacionalinis IV ir VIII ..., 2005*). It also appeared that the appyling of knowledge is performed worse than the demonstration of knowledge.

The goals of Lithuanian education require to focus teaching process on the students and their personal power development (*Pradinio ir pagrindinio ugdymo ..., 2008*). The teachers should develop a new methodology of the subject teaching - this methodology should correspond to the understanding of the present educational process based on the provisions of the interpretative didactics. In the order to improve the teaching of mathematics in the basic school, it is necessary to examine the actual facts of the educational situation constantly, to highlight the peculiarities of the methodology of teaching and learning mathematics and to search the features of the optimization in the teaching process. So it is appropriate to analyse the change of the teaching methodology, focusing more on the methods of active, inquiry-based teaching and learning.

Purpose and rationale

The article presents the results of the tests performed in 2004, 2006 and 2008. The statistical data of the first two studies was analysed on the aspects of the features and alteration of teaching methodology, which teachers apply in the V-X forms. Investigating teaching techniques of mathematics in the basic school, it was examined how often the teachers apply such modern teaching and learning activities as: the use of computer technology (ICT), the organisation of mathematics projects and excursions. It has also been interest in how often teachers use the elements of mathematics history in the lessons. The aims of this study were: to compare these results with the results of the formerly research, to sum up some results of earlier studies and to highlight some characteristics and peculiarities of teaching mathematics in the basic school.

According to the educational theory of constructionism the constructs of students' personal knowledge are determined on the activities of students learning process in the classroom (*Kolb*, 1984). Thus, the teaching methodology applied in the organization of students activities in the classroom, act on the students' learning process. As the various methods have different effectiveness of diverse mathematical competencies, it is important to combine the different teaching methods. As stated by R.I. Arends (1998), good teachers have the full repertory of the best teaching methods. They are not limited by their favorite ways and apply different methods, because lots of teaching methods are appropriate. The selection of the particular teaching method depends on the aim of teaching and the specific characteristics of the students group.

Organising the students' meaningful learning activities the teachers use teaching methods belonging to both groups: conventional and modern. There are different classification systems of traditional methods based on different classification criteria. For example, understanding, use and creative methods were allocated according to the level of cognitive activity; theoretical and practical methods according to the relationship between theory and practice; reproductive and interpretative methods according to the degree of creative activity (*Šiaučiukėnienė, 2006*).

The classification of mathematics teaching methods in the Lithuanian educational literature

The frequently used criteria of the classification of traditional methods in the educational Lithuanian literature is such as the opportunities' development of the independence and creativity in education (*Jovaiša, 2001; Šiaučiukėnienė, 2006*). According to this classification the informational, practical – operational and creative methods were assigned. The Lithuanian authors of mathematics methodical literature (*Drėgūnas, Rumšas, 1984; Ažubalis, 2008*) highlighted the mathematics teaching methods which can be classified in these three classes.

The *informational methods* are the system of the didactic actions by which the subjects' knowledge is conveyed for the students (*Jovaiša, 2001*). These methods can help to prepare students for the independent learning and do not require high creativity: storytelling, lecture, didactic conversation, demonstration, working with the manual. The Lithuanian authors of mathematics methodological literature (*Drėgūnas, Rumšas, 1984; Ažubalis, 2008*) noted that the explanation, school lectures, storytelling (for example, mathematics history), work on the book and other sources, demonstration methods can be used in the mathematics lessons effectively.

The *practical - operating methods* build up students' skills successfully, but they require much more self-sufficiency and creativity. They are: exercises, graphical, laboratory and practical work. It is recommended to use laboratory and practice (counting, measuring, plotting, model production) techniques in the mathematics lessons.

The level of the autonomy and creativity in is significantly higher, when teachers use *creative teaching methods: problem* (problem teaching and learning, problem discussion, problem solving, algorithm training, technical modelling); *heuristic* (heuristic conversation, logical proof, search); *investigative* (observation, experiment, research discussion). The methods of this class takes very important place in the repertoire of teaching methods by every qualified mathematics teacher. *Investigative (research) teaching methods* involve students into alleged or real scientific work. Using these techniques students are taught to explore the reality not only on the basis of logical reasoning and the use of literary sources, but also in the practical investigative work.

The ideas of inquiry – based teaching

The dissemination of the investigation methods in the global educational practice are associated with the originator of the pedagogical theory of pragmatism - J.Dewey's ideas of active teaching and learning (*Dewey, 1997; Teresevičienė, Adomaitienė, 2000*). According to the proponents of this theory, school education should be encouraged to seek answers to various questions of social life, to think creatively and to act innovatively. J.Dewey proposed an experimental training, when students solve problems while studying, experimenting and learning subjects, which are important to them. At the beginning of the last century the project method was developed through the ideas of activity-oriented learning.

Among proponents of the project method were listed W.H Kilpatrick and J.A. Stevenson (*Knoll, 1995; Kochhar, 2002*). They highlighted the essential features of the pragmatism educational theory, which distinguish the training ideas from so-called traditional learning at that time. The new learning is based on logical reasoning, but not on memory; the applicalble aspect of knowledge is emphasis; students solve the daily problems and construct their theoretical knowledge in this process to the contrast in the past, when they tried to apply in practice their theoretical knowledge stored in the classroom.

J. Dewey (1997) affirmed that the scientific exploration of the world directs students from passive learning to a higher level of thinking in which students learn: to raise the questions, to plan and design an investigation, to investigate, to formulate explanations and draw conclusions, to present findings and obtained results, to reflect findings and assure that the results are meaningful.

Through the research and experiments students learn to ask open questions, which often do not have unambiguous answers. When students analyze the results and demonstrate the findings, their open questions help to formulate reasonable explanations. Thus, answers to the raised questions, open the way to deeper exploration of the phenomenon or object understanding. In that way students acquire new knowledge, based on previously existing knowledge, experience and accomplished experiments.

Searching for the answer to the raised question and trying to solve the formulated project problems, students solve a number of tasks: to design an investigation, to choose adequate means and information sources, to collect information and necessary data, to analyze findings, to draw conclusions which are needed to explain and confirm decisions, to organise presentations.

In this activity students develop their critical thinking and creativity, problem solving skills, the abilities of information searching and analysis, the skills of questioning, inference, presentation, etc. Familiar with the studies and the results of their classmates, students have an opportunity to discuss the investigation, to highlight the similarities and differences, to share experience. In that way they can develop their capabilities of communication and collaboration. Scientific inquiry causes a change from traditional teaching practice to collaborative relationships between students and the teacher. Teachers become facilitators of students' creativity and model their scientific inquiry skills (*National Reseach Council...*, *2000*).

Mathematical projects and excursions as modern teaching methods. The elements of mathematics' history in the lessons

Nowadays the development of inquiry projects, based on the investigation, is considered as an effective learning activity. Projects and educational excursions (tours, trips, voyages, visiting museums or mathematical walks) are modern methods in teaching mathematics (*Cibulskaitė, 2001;*

Teresevičienė, Adomaitienė, 2000; Kochhar, 2002; Šiaučiukėnienė ir kt., 2006; Doering, 2006; Zollinger, Henderson, Atencio, 2007; Spijkerboer, 2008). It should be stated that a mathematical excursion is not identified in Lithuanian educational syllabus (*Pradinio ir pagrindinio ugdymo..., 2008*) as an independent teaching (learning) method. However, the programmes refer to the project work which has to guarantee the requirements of integration and help students to develop their learning provisions in mathematics. Projects can be prosecuted during mathematical excursions.

Fig.1: Golden section in geometry, architecture, art, nature and music

A principle of historicism is the usage of data of mathematics' history and presenting "the history of mathematics development, and introducing to the creators of mathematical science, their discoveries and investigational search". It can play an important role in this process (*Cibulskaitė, 2000, p. 61*). The importance of historical knowledge and comprehension of mathematics development in educating students' motivation in learning and promoting their interest in this discipline is emphasised in new syllabus (*Pradinio ir pagrindinio ugdymo..., 2008*). This is a productive way of teaching students to explore mathematical problems.

Example 1

The project *"Golden section in geometry, architecture, art, nature and music"* which connects mathematics with other curriculum subjects was created by students of the XIth form (ages 17-18). They examined the problem: *"Whether it is true that the number $\frac{1+\sqrt{5}}{2} \approx 1.618$ expresses the perfect (gold, divine) proportions of nature, human body and aesthetical creations of human beeings?"* (*Cibulskaitė, 2001*). Five small-scale scientific students' studies represented the theoretical foundations of the problem (Fig. 1). The students had to solve Euclids' task of dividing a line in the golden ratio. They were interested in theoretical elements of architecture, art, music and proportions of nature's objects. Historical information for searching of harmony of ancient philosophers and mathematicians, of Phidias, Leonardo Fibonacci, Luca Pacioli, Leonardo da Vinci and John Sebastian Bach works has been described in the students' articles. The project's results were originally presented to the school community - a five-part performance was shown.

Practical research work *"Golden section in our life"* has contributed a selection of other articles. This article was prepared by students of the VIIth form (ages 13-14). The problems were studied by those students: *"Whether people really prefer "golden" rectangles? Whether the proportions of paintings, packages and a human body are close to the golden section proportion?"* (Fig. 2). Some results surprised the students: 85 percent of surveyed people bodies' proportions do not deviate from the "golden section" in more than 10 percent; teenagers are "more proportional" than adults and women are "more proportional" than men.

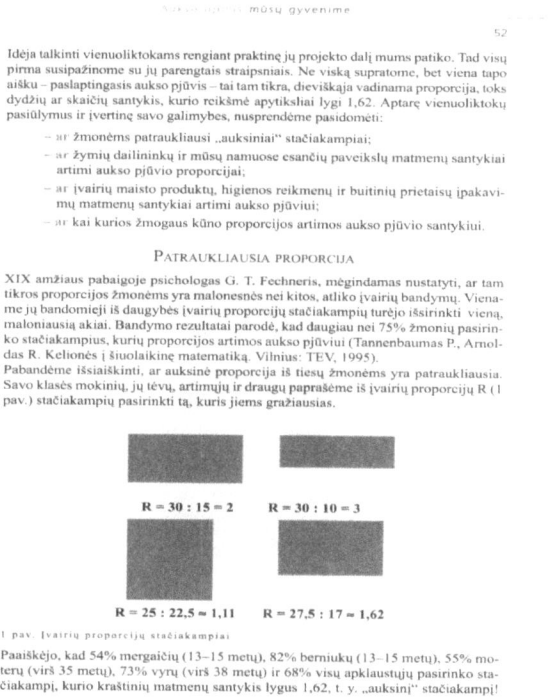

Fig. 2: Golden section in our life

Example 2

The textbook „*Matematika XXI a.*" (*Mathematics in the twenty-first century*) for the Vth form (*Cibulskaitė, 2005*) contains the following problem:
"The ancient Greek philosophers-sophists (IV-V century B.C.) contributed to the development of logic - a branch of mathematics. They argue the statement, called sophism: "Half-full is the same as half-empty. The two

sides compose the unit. Thus, the full is the same as empty". Why is this reasoning wrong?"

Another sophism presented for the students is: "The drawing shows two unequal rectangles made of equal polygons. One of them consists of 64 boxes, and the other of 65". The students are advised to find the error in their practical work. (Fig. 3).

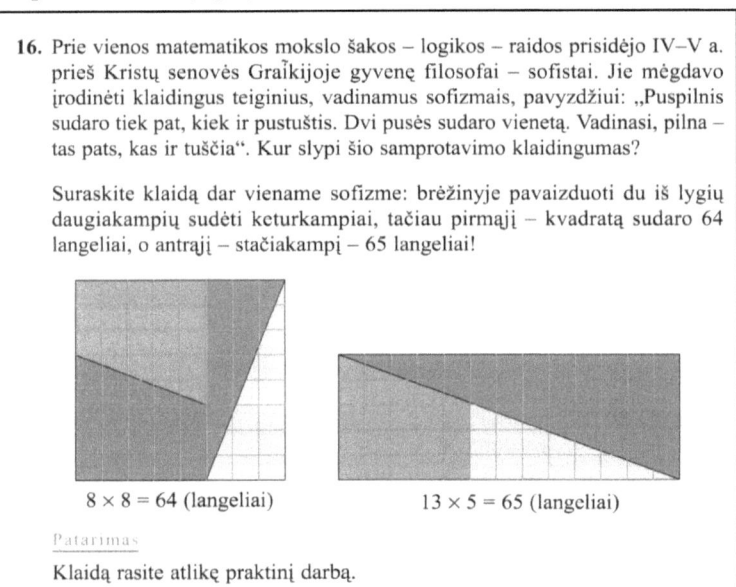

Fig.3. Sophistical problems

A mathematical excursion is a certain type of projects. The author called this training method as *Math walk* states: "The most important point of the mathematics walks is to learn to recognize mathematics in the world around us. As soon as something can be seen or calculated in realistic situations, it becomes clear how mathematics can be used. In mathematics walks there is plenty of scope for the use of knowledge and skills learnt in maths lessons" (*Spijkerboer, 2008*).

Example 3

The learning material for class V (relationship of measurement units, fractions, percentage, etc.) is revised when students are solving 21 tasks of the mathematical excursion in Vilnius Old Town (*Cibulskaitė, 2005*). Abilities to evaluate the dimensions of objects and identify geometric shapes and three-dimensional bodies are being trained in the environment. (Fig. 4, 5, 6).

✓ Netoliese – Aušros vartų g. namas Nr. 8. Tai XV a. gotikinis pastatas, papuoštas ornamentinėmis plytų juostomis bei Renesanso epochai būdingu sgrafitu.

7. *Kokias geometrines figūras įžiūrite sgrafito ornamentuose?*

✓ Paėjėję žemyn, iš kairės pamatysite buvusio Bazilijonų vienuolyno vartus. Vienuolyne veikė mokykla ir biblioteka, vėliau – spaustuvė.

✓ Einame toliau. Iš dešinės – vertinga ir unikali senamiesčio dalis, Subačiaus gatvė ir tolyn nusidriekęs Užupio rajonas. Čia išliko XV–XIX a. pastatų. Tarp jų – minėtoji bastėja.

Fig. 4: Mathematical excursion in Vilnius Old Town, task No.7

Let's stop at the house No.8 in Ausros vartu (Aushra gate) street. It is a Gothic building, decorated with an ornamental belt of bricks and sgraffito, which is characteristic of the Renaissance.

7. *What geometrical figures can you see in sgraffito ornaments?*

We are going towards the Town Hall. We can see St.Casimir's Church, which was built in 1604 – 1615, on the right side. It was the first building with cupola in Vilnius.

8. *Estimate the height of the Church's including cupola and choose the suitable answer:*

a) 20 m., b) 30 m., c) 40 m., d) 50 m.

✓ Einame Didžiąja gatve rotušės link. Iš dešinės – šv. Kazimiero bažnyčia, iš aplinkinių pastatų išsiskirianti dydžiu ir monumentalumu, nors jos kompozicija gana santūri ir paprasta. Tai pirmasis Vìlniaus statinys su kupolu.

8. *Nustatykite bažnyčios su kupolu aukštį, rinkdamiesi tinkamą atsakymą:*

a) 20 m; b) 30 m;

c) 40 m; d) 50 m.

✓ Priėjome Rõtušės aikštę – ji įsikūrė pagrindinių kelių į Trakùs, Ùkmergę, Pòlocką, Mẽdininkus ir Rūdininkus sankryžoje.

Čia senamiesčio širdis. Manoma, kad turgaus aikštė atsirado pre-

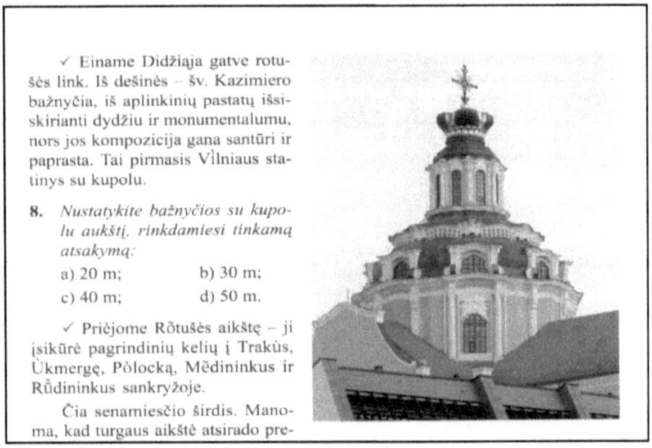

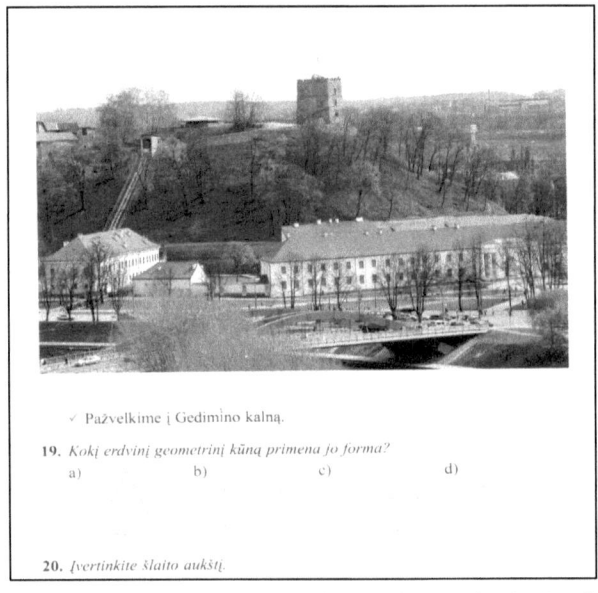

Fig. 5: Mathematical excursion in Vilnius Old Town, task No.8

✓ Pažvelkime į Gedimìno kalną.

19. *Kokį erdvinį geometrinį kūną primena jo forma?*

a) b) c) d)

20. *Įvertinkite šlaito aukštį.*

Fig. 6: Mathematical excursion in Vilnius Old Town, tasks No.19 and 20

Look at the Gediminas hill.

19. What dimensional geometrical figure does this hill remind of?

a) cone, b) cylinder, c) parallelepiped rectangular, d) cut-off cone.
20. Estimate the height of the hillside.

The ICT usage in the teaching/learning process

Modern information technologies used in educational process greatly increase the possibilities of conventional active methods. Information and communication technology (ICT) enables more effective use of all previously discussed methods (*Markauskaitė, 2000; Kraujutaitytė, Pečkaitis, 2003; Jonassen, and others, 2003*).

Statistical indicators show that the number of computer fall on 100 students of Lithuanian secondary education has grown rapidly during last five years (Fig. 7): from 4 school computers in 2003/2004 to 8.5 computers in 2008/2009 (LR Statistikos departamentas..., 2009).

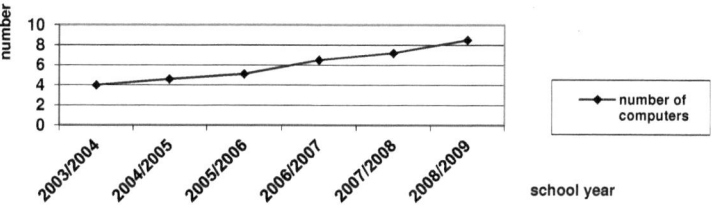

Fig. 7. The number of computer fall on 100 students of Lithuanian secondary schools

The data of National achievement test of IV and VIII forms carried out in 2007 (*Nacionalinio IV ir VIII klasių mokinių..., 2008*) shows that the number of VIII form students, who have got a computer at home, has increased from a little bit more than a half in 2003 to nearly two-thirds in 2005 and more than four-fifths in 2007. One third of VIII form students in 2005 and two thirds in 2007 indicated access to the internet.

Provided data suggests that the conditions of the use of ICT in the teaching/learning process at school and learning at home have recently improved.

Methodology

In order to establish the features and trends in change of methodology which teachers used in teaching/learning process at basic school over the past five years, a few tests have been carried out. While carrying out the research of the students of V - X forms (age 12 – 17) at basic and secondary schools in the cities, towns and rural areas, the number of students were interviewed: 912 in 2004 (response level 91 %), 1177 in 2006 (response level 95 %), 882 in 2008 (response level 86 %). Only fully completed questionnaires were used on the data analysis. According to the survey's geography and the overall number, the inquired students can represent all the students of the V-X forms of the republic. On the other hand, rather small groups of students of different forms were interviewed, so the results are used to highlight general trends and they are not generalized in the whole population.

The surveys' questinnaires, which were approved on the level of experts, allow testing not only mathematics teaching methods, but also testing some aspectcs of the relationship between students and teachers (*Cibulskaitė, 2003*). The sociodemographic and other characteristics of surveys' participants were included (gender, place of residence, school type, teachers' qualification, etc.). The questionnaire consisted of three groups of propositions. This article presents the rezults of the analysis of empirical data gathered from the students answers on a few questions only of the first group. These questions sought to determine whether the teachers organize the project work, use a computer in the classroom and at home, arrange mathematical excursions. In response to these questions students had to choose "yes" or "no". The empirical data was processed by using statistical quantitative (percentage frequencies and confidence interval with 95 % confidence), qualitative and comparative analysis.

Results

Mathematical projects

The positive responses of students on the question how often the teacher asked them to present their group or individual project, is given in Figure 8. Basing on the data obtained in the first study, it was possible to conclude that only up to a quarter (21 - 25 %) of the students of V and VI forms and up more than the third (30 - 36 %) of the senior students of VII - X forms were asked to present their projects.

The second study showed that the frequency of the projects organization in V - IX forms has increased: from two to three-fifths of students (39-62 %) presented their own or group projects (significant differences in the VI – VIII forms). Meanwhile, the frequency of the projects' organization in the X form was reduced: more than one third (36 %) of the students have delivered the projects previously and only a quarter (25 %) answered this question in the affirmative way at present.

The increase in project organization of a given period is related to the popularity of this activity. This kind of popularity possibly depended on the introduction of the profiled (humanistic – realistic) education and on the transition of the updated curriculum (LR *Bendrojo lavinimo ..., 2003*).

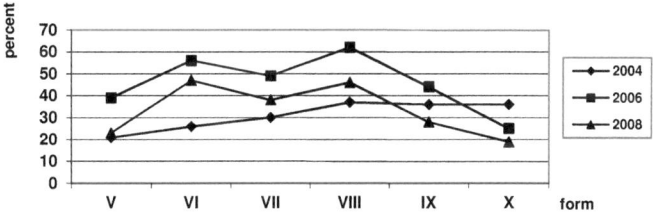

Fig.8: Positive answers of the students of V –X forms
on a question about project organisation (%)

The third study has revealed the project decrease in the frequency in all forms: only from less than one fifth to a half (23 - 47 %) of the students of V - X forms answered positively (significant differences in the V – IX forms). The trend of a downward of the project organization in the X form is distinct (at first 36 %, later 25 % and 19 % at last).

Recently the profiled education in the republic is rejected, the curriculum restructuring is completed, the schools convert to new educational programmes (*Pradinio ir pagrindinio ugdymo ..., 2008*), the mathematics textbooks of a new generation are used at schools nowadays, the requirements for the students' achievements in the VIII and X forms of the basic school are growing up, high quaity of education is expected from secondary schools. All these factors may determin the decline number of projects in basic schools: teachers can't spend much time on the projects sufficiently.

History of mathematics

Positive student responses to the question of how often the teacher familiarized students with the elements of mathematics history are presented in the Figure 9. The first study shows that between one quarter (26%) to more than one-third (36 - 38%) of the V form students and about half (49%) of the V-VIII forms students maintained familiar with the elements of mathematics history in the lessons. In the mathematics textbooks used at that time in the V-VIII forms there were many historical insertions, so the data shows that teachers did not use the potential of some books to interest students in mathematics enough.

The second study showed the increase of the number of students in all forms (except the VI and VIII forms), who stated they had heard stories of the history of mathematics in the classroom. The number of the students increased the least in the X forms (from 26 to 28%) and most in the V forms (from 49 to 55%), but the differences were not statistically significant. It is striking, but no statistically significant difference between the data recorded in the VI form. Recently only about one-fifth (22%) students stated they had heard stories of the history of mathematics in the classroom, although previously there were more than a third (37%).

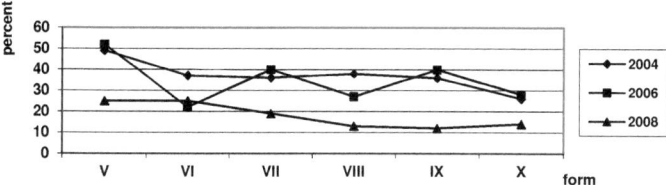

Fig. 9: Positive answers of the students of V –X forms on a question about history of mathematics during the lessons (%)

The third study revealed decrease of the frequency of the usage of the elements of mathematics history in the classroom in all forms, except the VI. Only one tenth to one fifth (12 - 19%) of the VII-X form students and a quarter (25%) of the V and VI form students answered this question in the affirmative (the differences in the V and IX forms were statistically significant). The data of the VI form compared with the data of the second survey increased slightly (from 22 to 25%), but the differences are not significant. Particularly evident downward trend of the usage of mathematics history in the VII - X forms. This indicates that the teachers does not take enough attention for the provisions of usage of mathematics history.

Some reasons for this may be the same as the downward trend of the above-mentioned projects and mathematical excursions organization. However, in order to highlight the specific causes of this phenomenon, new survey is required. Teachers of mathematics should be interviewed and survey would show why they rarely rely on mathematics teaching history items, and what help they need in order to establish program provisions.

Computer at home and in the mathematics classroom

Based on the data obtained in the first study, it was possible to conclude that the small number of teachers organize students' work at the computer:

only the tenth (6-12 %) of the students of V-VIII forms and one fifth (19 %) of the students of X form confirmed that they worked at the computer in the classroom and more than tenth (12-15 %) of the students of V-VI forms and approximately a quarter (22-25 %) of the students of VII - X forms performed the tasks at the computer at home.

The positive students' responses to the question of how often teachers organize their work at the computer in the classroom are presented in the Figure 10. The data results show that mathematics teachers applied the new technologies for training at that time rarely. They used the computer for teaching in IX-X forms of the basic school more often.

The second study showed that from the tenth to a quarter (9-26 %) of the students of V - X forms worked at the computer at school and from one fifth to nearly a half (20-46 %) of the students worked at the computer at home. In particular, the frequency of computer usage at home has increased in the VI and VIII forms (significant difference). The students of V and VI forms worked at the computer in the classroom more often (significant difference). This maybe related to the introduction of computer training course in the classroom startining whith VI forms.

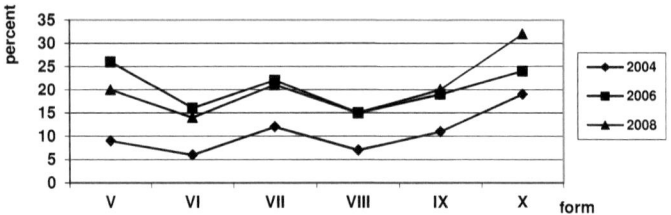

Fig. 10: Positive answers of the students of V –X forms on a question about computer using in the classroom (%)

The data of the third study showed a slight decrease in the frequency of computer usage in the classroom: up to one fifth (14 - 21 %) of the students of V – VII forms answered in the affirmative. It is observed that the stability of the data of VIII forms in which more than one tenth (15 %) of the students responded positively. The frequency of computer usage at school in IX - X forms has increased: the fifth to nearly the third (20 to 32 %) of the students answered positively. Differences in the X form are statistically significant.

When the question about the completed tasks at the PC at home was asked, from the fifth to more than a quarter (18 - 29 %) of the students of V – VII forms responded positively. Close to half (45 %) of the students of VIII forms and up to the third (28 - 33 %) of the students of IX - X forms responded positively too. The data is slightly different from the data received during the second study, however, it shows that teachers give computer tasks more often to the students of VIII – X forms.

Apparently this is related to growing computer literacy of the senior students as well as the increase of the computer training appliance and the greater opportunities to use a computer in the classroom and at home.

Mathematical excursions

Positive students responses to the question of how often the teachers alloted the tasks of mathematical excursions are presented in the Figure 11. In the first study it was indicated that the teachers gave tasks of mathematical excursions rather seldom: these tasks were carried out by more than one tenth (11-16 %) of the students of V, VIII and X forms and from the fifth to a quarter (22-26 %) of the students of VI, VII and IX forms. There were a number of tasks of mathematical excursions in mathematics textbooks of the V - VIII forms used at that time, so the results show that the teachers used the textbooks' opportunities for the active learning not well enough.

The second study showed that the students of the lower forms solved these tasks more often than senior ones as well as during the first study. From the tenth to more than one quarter (10-28 %) of the students of V and VI forms indicated that they carried out these tasks. The results of the students of V form increased and the differences were statistically significant. The indicators of VII-X forms were declined – only from the tenth to one fifth (10 to 21 %) of the students indicated that they had solved these tasks. Particularly, the number of students of IX form who performed mathematical tasks of excursions was reduced (significant difference).

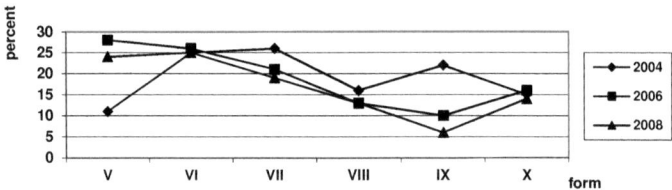

Fig. 11: Positive answers of the students of V –X forms on a question about the tasks during mathematics excursions (%)

The third study revealed the decrease of the frequency of mathematical excursions assigned to all forms: only up to a quarter (6 - 25 %) of the students of V - X forms have responded positively (although differences are not statistically significant). The trend of the downward of the mathematical excursions organization in IX – X forms is especially evident. This means that teachers use the possibilities of these tasks increasingly less in education of students' skills to practically apply knowledge in mathematics, link them with knowledge in other disciplines and express the importance of mathematics in society's life. The reasons of this phenomenon may be the same as the above-mentioned ones of the projects implementation.

Conclusion and discussions

The conspicuous tendencies are that mathematics teachers of V-X forms use a computer for teaching in the classroom and at home more often than before and especially in the senior forms of some Lithuanian basic schools. Also, significantly less than a few years ago, they used such modern active teaching/learning method as mathematical projects. As well the teachers also rarely used the elements of the mathematics history and alloted the tasks of mathematical excursions and significantly less often used this activity in senior forms of the basic school recently.

The results of the first two studies carried out in 2004 and 2006, showed that the teachers applied active teaching and learning methods (projects and mathematical excursions) insufficiently, though the textbooks for V - VIII forms used at that time were planned ahead. When the mathematics programme was changed in 2003, the new textbooks have been used in the basic school since 2005. The third study in 2008 showed that the teachers started using these learning activities even more seldom.

The textbook remains the basic form of teaching facility at present (*Nacionalinis IV ir VIII klasiu mokiniu...*, *2008*). So as the updated mathematics curriculum is valid from the year 2008 it is necessary to prepare new mathematics textbooks. The authors should think over the active training tasks deeply.

The recommendations based on survey results could be such as: mathematics teacher training institutions should focus on students ability to create modern teaching and learning methodology more directly, to master research and inquiry based methods. It is worth to teach students to apply the didactic principles of mathematics and encourage the uptake and usage of active teaching and learning methods and new modern technologies during the courses of mathematical didactics.

References

Ažubalis, A. (2008). *Logika ir mokyklinė matematika. Monografija.* [2010-12-27]. www.lka.lt/EasyAdmin/sys/Logika_internetui_1.pdf

Arends, R.I. (1998). *Mokomės mokyti*. Vilnius: Margi raštai.

Cibulskaitė, N. (2000). Matematikos mokymo(si) humanizavimas V pagrindinės mokyklos klasėje: daktaro disertacija. Vilnius: VPU.

Cibulskaitė, N. (2001). Aukso pjūvis geometrijoje, architektūroje, dailėje, gamtoje ir muzikoje. Vilnius: Kronta.

Cibulskaitė, N. (2003). Nūdienos matematikos mokymo XII klasėje metodikos ypatybės. *Lietuvos matematikos rinkinys*, spec.nr. T.43, V: MA MII, psl. 330 - 334.

Cibulskaitė, N. (2005). *Matematika XXI a.* Vadovėlis V klasei. Vilnius: Kronta.

Dewey, J. (1997). *Experience and Education*. New York: Simon & Schuster.

Drėgūnas, V., Rumšas, P. (1984). *Bendroji matematikos mokymo metodika*. Vilnius: Mokslas.

Jonassen, D., Howsland, J., Moore, J., Mazza, R. (2003). *Learning to solve problems with technology*. Englewood Cliffs, NJ: Prentice Hall.

Doering, A. (2006). Adventure Learning: Transformative hybrid online education. *Distance Education*, Vol. 27, N.2, pp.197-215.

Jovaiša, L. (2001). *Edukologijos pradmenys*. Šiauliai: Šiaulių universiteto leidykla.

Knoll, M. (1995). The Project Method: Its Origin and International Influence. In: *Progressive Education across the Continents. A Handbook*. Ed. Volker Lenhart and Hermann Röhrs. New York: Lang.

Kolb, D. A. (1984). *Experiental learning: Experience as the source of learning and development*. Engelwood Cliffs, New York: Prentice Hall.

Kochhar, S.K. (2002). *Methods and Techniques of Teaching*. New Delhi: Sterling Publishers Private Limited.

Kraujutaitytė, L., Pečkaitis, J. (2003). Nuotolinių studijų organizavimas: strategijos ir technologijos. Monografija. Vilnius: LTU Leidybos centras.

Lietuvos bendrojo lavinimo mokyklos bendrosios programos ir išsilavinimo standartai. I-X klasės. (2003). Vilnius: LR ŠMM.

Markauskaitė, L. (2000). Informacijos ir komunikacijos technologijos integravimo į ugdymą krypčių analizė. *Informatika*, vol.2(36).

Nacionalinis VIII klasės mokinių pasiekimų tyrimas. 2003 metai. Dalykinė ataskaita. (2003). Vilnius: LR ŠMM.

Nacionalinis VI ir X klasių mokinių pasiekimų tyrimas. Dalykinė ataskaita'2004. (2004). Vilnius: LR ŠMM.

Nacionalinis IV ir VIII klasių mokinių pasiekimų tyrimas. Dalykinė ataskaita. (2005).Vilnius: LR ŠMM.

Nacionalinis IV ir VIII klasių mokinių pasiekimų tyrimas. 2007 metai. Apžvalga. (2008). Vilnius: LR ŠMM.

National Research Council. (2000). Inquiry and The National Science Educational Standards: A Guide for Teaching and Learning. Washington, DC: National Academy Press.

Pradinio ir pagrindinio ugdymo bendrosios programos. (2008). Vilnius: LR ŠMM. [2010 - 01-14]. <http://www.pedagogika.lt/index.php?-469374926>

Statistikos departamentas prie Lietuvos Respublikos Vyriausybės. [2010 - 01-14]. <http://db1.stat.gov.lt/statbank/default.asp?w=1024>

Spijkerboer L. (2008). Designing mathematics walks. In: *Supporting Independent Thinking Through Mathematical Education.* Wydawnictwo Uniwersytetu Rzeszowskiego. 2008. [2010-12-12]. <www.cme.rzeszow.pl /pdf/part3.pdf >

Šiaučiukėnienė, L. ir kt. (2006). Šiuolaikinės didaktikos pagrindai. Kaunas: Technologija.

Tarptautinis matematikos ir gamtos mokslų tyrimas. Ataskaita. Matematika 8 klase. 2007. (2008). Vilnius: NEC.

Teresevičienė, M., Adomaitienė, J. (2000). Projektai mokymo(si) procese. Mokomoji knyga. Kaunas: VDU.

Zollinger Henderson, T., Atencio, DJ. (2007). Integration of Play, Learning and Experience: What Museums Afford Young Visitors. *Early Childhood Education Journal*, 35: 245-251.

Introduction of a historical perspective in physics secondary school: replication and use of Galileo's experiments between theatre and modelisation

Stéphane Le Gars

Centre François Viète, Université de Nantes, Nantes, France.

ABSTRACT: The introduction of a historical perspective in physics teaching is, as historians of science well know, a difficult question. Indeed, a simple evocation of a past scientist and of his contributions to the progress of science is not constructive and brings no meanings. So, can history of science participate to the construction of knowledge for pupils in secondary school ? This paper wants to illustrate an example where history of science, physics teaching and drama session were articulated to understand and show the experimental work of Galileo.

This project was an interdisciplinary one, based on the collaboration between pupils enrolled in drama session and pupils enrolled in scientific course (named "Mesures Physiques Informatisées"). The theatre point of view was elaborated on the problematic of telling science on stage. Imitating the form of the dialog used by Galileo, pupils have been invited to adopt the same form, for being initiated to an important element of the development of science, I want to say the controversy. But crucial questions have soon emerged: what kind of scientific knowledge can be treated, and how can it be treated ? How to show or make science inside a theatre ? How science can be integrated to a drama session ? It has rapidly appeared that previous experiences of science theatre realized by professional directors put the emphasis on the use of scientific objects on stage: these objects were thought to show that science is a human production, and they allowed the clarity of the play.

So, the aims of this project were varied: they were pedagogical, artistic and scientific. First of all, we tried to build and use scientific objects to make an artistic sense of them. The "scientific" pupils were invited to replicate an inclined plan and a refractor, as Galileo has described them: the instruments were therefore integrated to the play and showed to the public in an aesthetical perspective. For these pupils, a deep scientific understanding was expected, because "experimenting is creative work of the mind and the hands"[1]. The replication has, in fact, many virtues: it permits to gain a scientific knowledge, to develop handful skills, to historicize and contextualize the scientific progress: "rebuilding and redoing makes it possible to understand scientific work and its outcomes in a historical context through an intellectual as well as a sensual and aesthetical experience"[2]. Furthermore, the use of ICT resources renew the problematic between teaching and history of science: replicating historical experiments by the means of specific modelization soft wares and with the taking of video offers to the students a way of thinking about precision of measures and the induction of physical laws.

Secondly, the show was used to render science intelligible, and to tell science to the general public: the play was, indeed, a compilation of primary and secondary texts. Tselfes and Paroussi have recently questioned the link between the theatrical, narrative form and the scientific way of thought: for them, "science making" procedures can be represented theatrically, and "the closer the scientific ideas' sources are to the scientific creation's human and social dimension, the more narrative their structure is and more complete these approaches are from a literary and/or aesthetic point of view, the more chances there are for the students to gain more elaborated forms of understanding."[3]

Then, controversy was showed to be a necessary way in scientific development, and multiple style of disputes were studied: the polemic, the dramatic and the oratorical ones. In a teaching point of view, it has often appeared that the use of controversy in a pedagogical form could be a "good way to deconstruct the inductive view of the knowledge building process and to discuss with the students the role of experimentation in the construction of science."[4]

In this paper, we will conclude with a methodological reflexion. Indeed, in our theatrical/scientific project, students were separated between "scientists" and "artists". This is a methodological limit of our interdisciplinary project: how can every student benefit of replication and historical work, combined with the use of ICT resources, if a traditional separation between scientific and literary courses is first organized ?

The first step of this interdisciplinary project has consisted in an astronomical observation getting together "artist" and "scientific" pupils. This permitted to create a common "history" for the two groups of pupils (to go against the strong specialization of the french cursus in secondary school), and to introduce them to instruments such as reflectors and refractors. Pupils took this opportunity to observe the same astronomical objects than Galileo until 1610 (the moon and the satellites of Jupiter), and to photograph those objects by the means of a webcam (these images were then projected during the play to show what Galileo saw in his refractor).

After this first encounter, the two groups of pupils worked independently from each other: "artist" pupils studied and chose texts to create the play, and "scientific" pupils studied the same texts in the aim of building some instruments that could be integrated in the play. Three instruments were first thought to be integrated to the play: a refractor, an inclined plan and a pendulum. The latter one was rejected after, for technical reasons due to the impossibility of fixing it on the stage.

The construction of the refractor was based on the explanation given by Galileo in *Sidereus Nuncius:*

> "First I prepared a tube of lead, at the ends of which I fitted two glass lenses, both plane on one side while on the other side one was spherically convex and the other concave. Then placing my eye near the concave lens I perceived objects satisfactorily large and near, for they appeared three times closer and nine times larger than when seen with the naked eye alone. Next I constructed another one, more accurate, which represented objects as enlarged more than sixty times. Finally, sparing neither labor nor expense, I succeeded in constructing for myself so excellent an instrument that objects seen by means of it appeared nearly one thousand times larger and over thirty times closer than when regarded with our natural vision."[5]

The construction of the telescope and of its stand was then assured with the aid of indications given by a French review linking astronomers and teachers, *Les Cahiers Clairaut* [6]:

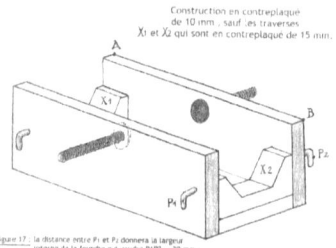 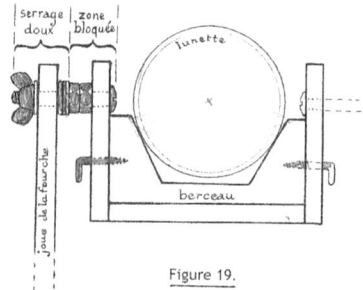

Figure 19.

The construction and replication of an inclined plan was more linked with modelization and permitted for pupils to think about the empirical method of Galileo. The basis of this work was made with a text of Galileo from the *Discorsi*:

> "A piece of wooden moulding or scantling, about 12 cubits long, half a cubit wide, and three finger-breadths thick, was taken; on its edge was cut a channel a little more than one finger in breadth; having made this groove very straight, smooth, and polished, and having lined it with parchment, also as smooth and polished as possible, we rolled along it a hard, smooth, and very round bronze ball. Having placed this board in a sloping position, by lifting one end some one or two cubits above the other, we rolled the ball, as I was just saying, along the channel, noting, in a manner presently to be described, the time required to make the descent. We repeated this experiment more than once in order to measure the time with an accuracy such that the deviation between two observations never exceeded one-tenth of a pulse-beat. Having performed this operation and having assured ourselves of its reliability, we now rolled the ball only one-quarter the length of the channel; and having measured the time of its descent, we found it precisely one-half of the former."[7]

It was decided to construct this inclined plane to incorporate it to the play. As Galileo did, time was measured with bells, making noise when a ball was hurting them: Actors "pupils" would use it on stage to illustrate the scientific context of Galileo's works. But, to the contrary of Galileo, it was decided to find the positions of the bells permitting to hear regular tinkling and not an accelerated one. So, a mathematical modelization was necessary to determine with precision where to place the bells on the inclined plan. A 3m-long inclined plan was built, with a metallic rail fixed on it, adapted to the length of the ball. Multiple parameters were listed and experimented:

the material constituting the ball, its mass, its diameter, the angle of inclination for instance. Various soft wares were then used to infer the mathematical law of the falling ball: Vidcap32 for the video acquisition, Avimeca for the video exploitation, and Regressi for the modelization of the positioning of the ball on the inclined plan.

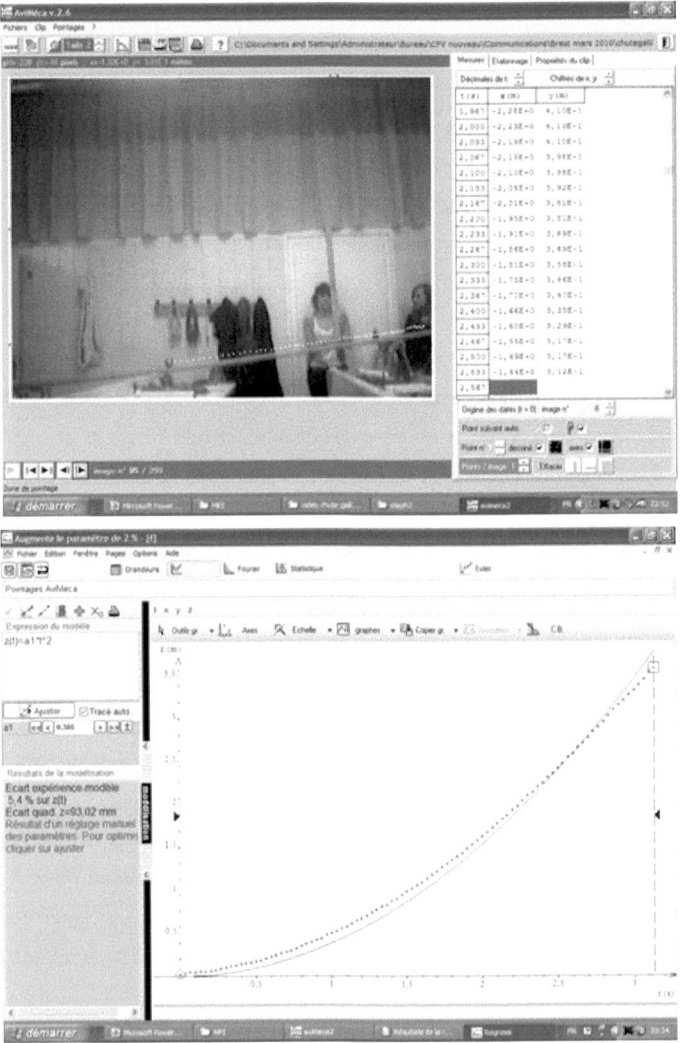

After various experiments, modelizations gave way to discussions about uncertainty, precision and the possibility of ratify the measures or not. Those discussions were based on the difference between the experience and the mathematical model that was chosen on the one hand, and on the criterion that the teacher imposed to the pupils to ratify the model (5%). Those discussions were extremely fruitful to show the relative partiality of scientific induction, and the emergence of polemics or controversies that can emerge to give and end to an experience: this exemplifies the fact that scientific creation is a social process.

Pupils chose a law of falling bodies that was: $z(t) = a.t^2$, but discussions emerged when some groups found that a lower difference between experience and model was obtained if the power of t (time) was not 2. Historical and epistemological problems could raise: is scientific progress limited by technological constraints depending on the historical context ? Are physical laws right if they are the simplest ones ? Do physical laws give us a true image of the universe ? What kind of beliefs did these laws support Galileo's works and interpretations ?

The second aim of the project was to render science intelligible, and to tell science to the general public. The play was created by mixing various texts: *La Vie de Galilée*, of Bertolt Brecht, the *Dialogo* and the *Sidereus Nuncius*, of Galileo, above all. The theatrical work was then realized during three distinct periods. During the first one, "artist" pupils had to realize exercises dedicated to the learning of theatrical pratice, in close connection to scientific practice or scientific symbolization in everyone's mind. For instance, pupils were questioned about the style of language that is used by the scientists (they were led to utter theorems), about the clothes that kind symbolize scientists best (the famous white coat). The use of masks permitted as well to initiate pupils to the idea of a scientific persona, the public and social dimension of the scientist: in the case of Galileo, the use of masks allowed a thinking about this idea of a scientific persona, and about the ambiguity between mask and persona, between public and intimacy. Finally, during this first step, improvisation was privileged, above all in the aim of twisted objects.

In a second step, those improvisations were used to the work of stage compositions, incited by scientific images (the ones of the moon and of Jupiter and its satellites taken at the beginning of the project), and by scientific objects built by "scientific" pupils. From texts of Brecht, Galileo or Pascal, various acts could be written and composed.

In the end, during a third moment, a 30 min theatrical creation raised, with the aid of a professional actor, musical moments were integrated, and scientific images and objects took a place that was the concretization of scientific and theatrical work all at once.

In a last part, it seems necessary to bring up the problem of controversy: this scientific aspect has been a central point of the pedagogical method employed during this project. It's first an important tool for the historians of science. As Pestre writes: "[controversy] constitutes an *irrepleaceable methodological tool* for historians, philosophers and sociologists of sciences, because it permits to question the epistemological problem in a new way: the privilege given to meticulous support of the construction of theories and experimental facts, as well as of the arguments that the various players exchange about them, allows to show the openness of the constructions and proofs that are proposed, and then to see how ones are led to express what constitutes the agreement"[8].

So, this project is based on privileged methods used by historians of science today, and this at two levels: improvisations and theatrical learnings were based on disputes, controversies and debates, by the language and by the bodies as well. But controversy was as well used during the scientific work of the "scientists" pupils as we have seen above: all at once "the experiments and the texts put the students in conflict with their first vision of science"[9]. In this project, only an inquiry based science teaching, using ICT resources, has offered the occasion to debate between epistemological positions at the end of the experimental process: creative process has led to controversies and has participated to change a first vision of science based on induction and not polemics and debates.

As a conclusion, it's important to notice the limits of our project: indeed, we can see that, because of organizational constraints, "scientists" and "artists" pupils were separated. Is this relevant and pedagogically efficient ? In their article about science and theatre education, Tselfes and Paroussi bring up the problem of separation between artistic and scientific education in our societies. They demonstrate that the teaching and learning of scientific ideas can coexist with theatrical expression, if teaching, based on narrative structures, takes the form of lively processes of science making, rather than being an account only of finished science (Tselfes and Paroussi do their own Bruner's ideas)[10]. In this way, it is the problem of creation that is pointed out: "The hypothesis that can be seen as asserting itself is that the separation between scientific and artistic activities in the field of education arises from the fact that, as a rule, scientific activities are not creative in the way artistic ones are."[11] Tselfes and Paroussi show that if the official scientific activity is based on creative processes such as the constructions of hypothesis or models, in schools, pupils are rarely offered the possibility of scientific creation. In this way, "the field of theatrical expression is a field that involves students creatively with the construction of a more or less artistically successful theatrical study. This creative activity can call for the use of scientific/epistemological ideas (which constitute dramatic functions, the performance's holistic referent, etc.) in ways that require at least one form of understanding of them."[12]

If we place the problem of creation as a central problem of our project, it appears that the separation between "scientists" and "artists" pupils is not a real problem. The use of ICT resources and methods for Inquiry Based Science Teaching allows the communication between the scientific and the artistic parts, by the means of images and objects integrated in the play. Scientific and artistic learnings are based on creative processes, and rejoin themselves with the use of numerical modelizations, new imagery acquisitions processes, historical and epistemological materials, digitized images projections. In this point of view, the play and the lab appear as boundary places where emotional ways are used to show that science is a social construction, and that science and theatre are two creative processes that must be intertwined: theatrical projects are a privileged place to show the necessity of questioning the links between science and society.

References

Braga, M.; Guerra, A.; Reis José, C. (2010) "The role of historical-philosophical controversies in teaching science: the debate between Biot and Ampère", *Science and Education*, online first, 21 october 2010.

Brisville, J-C. (1986), L'Entretien de M. Descartes avec M. Pascal Le Jeune, Paris, Actes Sud.

Brecht, B. (1990), *La Vie de Galilée*, Paris, L'Arche.

Dewinter, C. (1999), Lectures d'une œuvre. La Vie de Galilée de Bertolt Brecht, Paris, Editions du Temps.

Galileo Galilei (1992), *Dialogue sur les deux grands systèmes du monde*, Paris, Editions du Seuil.

Galilée (1995), Discours concernant deux sciences nouvelles, Paris, Puf.

Galilée (1989), Le Message céleste. Traduction complète du Latin en Français avec des notes par Jean Peyroux, Paris, Librairie Blanchard.

Höttecke, D. (2000), "How and what can we learn from replicating historical experiments ? A case study.", *Science and Education*, 2000, 9, 343-362.

Pestre, D. (2006), *Introduction aux science studies*, Paris, La Découverte.

Rhys Morus, I. (2006), "Seeing and believing science", *Isis*, 2006, 97, 101-110.

Tselfes V.; Paroussi A. (2009), "Science and Theatre Education: A Cross-disciplinary Approach of Scientific Ideas Adressed to Student Teachers of Childhood Education", *Science and Education*, 2009, 18, 1115-1134.

Valmer M. (2005), *Le théâtre de sciences*, Paris, CNRS Editions.

[1] Höttecke Dietmar, "How and what can we learn from replicating historical experiments ? A case study.", *Science and Education*, 2000, 9, p.358.
[2] Ibid.
[3] Tselfes Vasilis, Paroussi Antigoni, "Science and Theatre Education: A Cross-disciplinary Approach of Scientific Ideas Adressed to Student Teachers of Childhood Education", *Science and Education*, 2009, 18, p.1132.
[4] Braga Marco, Guerra Andreia, Reis José Claudio, "The role of historical-philosophical controversies in teaching sciences: the debate between Biot and Ampère", *Science and Education*, online first, 21 october 2010.
5 With pupils, we used this edition: GALILEE, Le Message céleste, Traduction de Jean Peyroux, Paris, Blanchard, 1989 [1610]. The translation we utilized is available on this site:
http://www.bard.edu/admission/forms/pdfs/galileo.pdf
[6] Voir *Les Cahiers Clairaut*, n°123, automne 2008.
[7] With pupils, we used this edition: GALILEE, *Discours et démonstrations mathématiques concernant deux sciences nouvelles*, Paris, PUF, 1995 [1638], p.144 [§ 213]. The translation we utilized is available on this site:
 http://galileoandeinstein.physics.virginia.edu/tns_draft/tns_160to243.html
[8] Dominique Pestre, *Introduction aux science studies*, Paris, La Découverte, 2006, p.26.
[9] Braga Marco, Guerra Andreia, Reis José Claudio, *op.cit.*
[10] Tselfes Vasilis, Paroussi Antigoni, "Science and Theatre Education: A Cross-disciplinary Approach of Scientific Ideas Adressed to Student Teachers of Childhood Education", *Science and Education*, 2009, 18, p.1119.
[11] *Ibid.*, p.1132
[12] *Ibid.*

Suggestions for Introducing the History of Chinese Technology into Education

Carles Puig-Pla

Centre de Recerca per a la Història de la Tècnica, ETSEIB, Universitat Politècnica de Catalunya; carles.puig@upc.edu

ABSTRACT: From the 2nd century BD to the 16th century, the capacity to apply knowledge of nature to practical purposes was really more effective in the oriental Asian culture than in Western European culture. However, Chinese technology and the inventions achieved by this millennial civilization are still unknown to the majority of the population in the West. The experience acquired with university students enables the author to gather information about how to introduce the scientific and technological innovations of Ancient China to students in the main fields of Chinese innovations. This experience suggests that it would be extremely useful to introduce some aspects of Chinese history and culture prior to specific knowledge of science or technology. Furthermore, new and readily available ICT resources can provide very helpful teaching tools.

Introduction

This work deals with the author's experience in teaching the history of technology of ancient China at the Polytechnic University of Barcelona (Puig-Pla, 2007). The author teaches an elective subject called "Culture and Technology in Ancient China". This program consists of two distinct parts: the first concerns cultural aspects, while the second is of a more technical nature. In general, the use of new ICT resources has a significant effect in education and epistemology, and the history of science and technology (EHST). As explained below, this is also the case for the introduction of the History of Chinese technology.

The first part of the program: general and cultural aspects

Some issues related to China today are first considered; basically, population and territory in the current People's Republic of China. China is a state occupying a large territory, the third largest in the world. Geomorphologic characteristics, climate, major crops, the ethnic majority, autonomous regions, climatic contrasts, rainfall patterns, nationalities, etc. all come under study.

It is important to point out the extent of China's provinces in comparison with other countries occupying a similar area. No less important is the large number of Chinese inhabitants, more than one thousand three hundred million. The population of some provinces may be compared with the population of entire countries; for example, some provinces such as Guangdong are as highly-populated as Mexico, a country with the same population, or the province of Yunnan, which has the same population as Spain. (For more on the current People's Republic of China, see 1.1 in the ICT Resources Annex).

Secondly, it is necessary to consider some of the cultural characteristics of China, especially those that are different from the West. Apart from the many sinitic languages spoken in China, particular emphasis is placed on the common language, Putonghua or Mandarin, which is a monosyllabic language with 4 tones whose problematic feature is its multiplicity of homophones (see ICT Resources Annex, 2.1 time interval 1:25 - 2:26).

Some reference to Chinese writing is also required. Because of the desire of philosophers (who nowadays would be called scientists) to find a system to ensure accurate communication, Leibniz himself became interested in Chinese writing (Doroudi, 2007). In fact, although Chinese people write in the same way, not all of them speak in the same way. At this point it is necessary to give a brief revision of Chinese characters, which may help students to understand Chinese culture more deeply and more easily (Fazzioli, 1987). It is instructive to consider the basics of written Chinese; the special elements of Chinese characters (radical strokes or keys); the way a Chinese dictionary is compiled and used; the problems concerning the different Chinese language transcription systems and the existence of Pinyin (the official phonetic transcription to the Latin alphabet, created in 1979 to avoid confusion); the evolution of some pictograms, and so on.

Students are particularly attracted to this topic (the Chinese language and writing), and every year there is an active involvement of students when the teacher explains this topic. The many questions asked by students reflect their interest in Chinese culture, which is one of the initial goals for situating students in this different context.

This section devoted to Chinese culture can be extended to other cultural areas covering artistic activities such as Painting, Calligraphy and Poetry; the Peking Opera or Traditional Chinese Medicine, among others.

The course then leaves the present era and addresses the past through the exposition of some basic events in Chinese History. Whereas in Western history centuries are used to measure periods of time, in China the most usual method is dynastic chronology. The division of Chinese History into dynastic periods makes more sense than the division by centuries used in the West (Fairbank 1996, p. 72). It is important for students to learn and remember the chronology of the major dynasties. Finally, the first part of the program ends with Traditional Chinese thought (Confucianism, Taoism, Buddhism...) and its tendency to syncretism (Cheng 1985, p. 34). This cultural and historical background will be relevant for understanding and placing in context many Chinese inventions. For example, Astronomy or the origins of printing can be associated with Confucianism; Gunpowder with Taoism or Woodcut and Xylography with Buddhism.

Sections 1 (Population and Territory) to 4 (Traditional Chinese Thought) of the "ICT Resources Annex" contain a large amount of online information about the first part of the program. The use of these digital resources greatly facilitates the understanding of historical and cultural context of scientific and technological production in China. Students enjoy free access to these resources and the teacher can choose specific snippets for display and discussion in the classroom.

The second part of the program: Chinese Technological innovations

The second part of our Program Proposal deals with the main fields of "Chinese scientific & technical innovations". These fields can be generally classified into nine groups.

    2.1 Paper and Printing
    2.2 Proto-chemistry (chemistry of explosives: gunpowder)
    2.3 Magnetic Physics (compass)
    2.4 Technology of Iron and steel
    2.5 The use of animal power
    2.6 Navigation and nautical inventions
    2.7 Astronomy and astronomical instruments
    2.8 Mechanical Clock
    2.9 Domestic Technology

In addition, students are provided with a list of topics for an assignment to be completed alone or in a group. The list of topics is determined by the program, but remains open to new proposals and suggestions from students themselves. The list covers topics related to epistemology and the history of science and technology (EHST). The students are required to choose one such topic as the basis of a research paper. For example, the topics concern *military technology, civil engineering, the old bridges, the techniques of the alchemists, Grand Canal, textile technology,* and so on. Should students express a particular interest in the Chinese world, they can if they wish propose a new topic related to that interest: from *Traditional calligraphy techniques* to cultural or technological aspects that are closer to modern history, such as *The construction of the hydroelectric Three Gorges Dam* or *How business is done in China*.

The teacher provides guidance on various sources (reprints; texts available on-line; specific resources available in the University Library, etc) to students, who usually have one month and a half to prepare their research assignments. After this, they are required to give an oral presentation of their work in the classroom. At the end of each oral presentation, their work and the topic are discussed in a debate with other students in a seminar session, the purpose of which is to assess the formal presentation (about 20 minutes); the critical acumen in the choice or comparison of sources, especially if Internet and Communication Tools (ICT) are employed; the work in groups, if necessary, and students' ability in debate and argument subsequent to presentation

One important classical source for study is *"Science & Technology in China"* (1954-2008), which arose from the extraordinary work of Joseph Needham (1900-1995) and the research carried out at the Needham Research Institute in Cambridge. Needham promoted the development of this monumental work "Science & Technology in China"; 27 volumes of which have been published so far. Apart from Needham, other historians of Science have investigated the history of Science and Technology in China, such as Colin Ronan (1993), Jean-Claude Martzloff (2006), Robert Temple (1986), Christopher Cullen (1996), Karine Chemla and Guo Shuchun (2005), among others.

Based on these and other works, resources are currently available online that can be used to teach the history of Science and Technology in Ancient China. It is not difficult to access these resources on the Internet and to prepare an ICT resource list for classes. In this regard, section 5 of the ICT Resources Annex is devoted specifically to ancient Chinese inventions. (See the ICT Resources Annex, 5.1 to 5.10).

Without entering into detail, I now provide a brief outline about the second part of the course. Normally, the four great inventions of Ancient China: *papermaking, printing, gunpowder* and the *compass* are first introduced. Inventions are regarded as technologies first developed in China and as such do not include foreign technologies acquired by the Chinese through contact with other peoples or countries. (For more about the four great inventions, see 5.1 to 5.4 in the ICT Resources Annex).

Thus, this second part of the course begins with the study of *Baqiao* paper (unearthed in Xi'an, Shaanxi), a plant fiber paper mainly made from hemp and bast fibers and dating from the Western Han Dynasty (2nd – 1st centuries BC). Subsequent improvements were made, such as those by Cai Lun, a eunuch at the Court in the Eastern Han Dynasty, who improved the method of manufacture, or those introduced by Zuo Bo, an expert in the manufacture of paper in the Late Eastern Han Dynasty. In this regard, the program also includes the study of some relevant achievements in paper technology: Xuancheng Paper, made from the bark of sandalwood and rice straw, which was highly hygroscopic, or the use of different fibers (mulberry tree, bamboo…), according to the areas where they were manufactured (Zhuang, 1980, pp. 67-76).

As regards printing, it is necessary clarify and distinguish between *Woodcut* or *Xylography* and Typography. Before introducing Woodblock printing, the origins of printing techniques are shown: Seals and tracing paper written in bas-relief on the stones for copying official Confucian texts in order to pass official state exams. Information is provided on a very old Woodblock printing text (Buddhist) from 636 (*Nüxe* or biography of women) (Zhuang, 1980, pp. 45-46). The Diamond Sutra (868), a surviving scroll of a Buddhist text, is one of the oldest printed woodblocks in existence. Woodcut was already fully developed in the 9th century, and included calendars, literary works and Buddhist texts. The first official publications appeared in the 10th century, and in the 12th century multicoloured prints were published in China.

Typography was introduced in the Northern Song Dynasty by Bi Sheng (? –1051), who, according to a contemporary source, invented movable types about 1041-1048. Subsequently, important improvements were made, especially the revolving table type case with individual movable type characters

(Liu & Zheng 1989, pp.75-83). This was an achievement of Wang Zhen (1297-98).

Obviously the course content also includes proto-chemistry and the invention of gunpowder and its applications, as well as the first alchemist experiences of *fangshi*.

The earliest known written description of the formula for gunpowder, which was written and compiled by Zeng Gongliang (1040), is highlighted. Naturally, sophisticated weapons using gunpowder, such as a flamethrower in early 10th century, rockets (11th century), portable rocket launchers (14th century) or a rocket battery on wheelbarrows (17th century), among others, are also studied. The course further addresses "how and when" Chinese technological discoveries and inventions reached the West. (See ICT Resources Annex 5.8, part 1/5 for gunpowder).

The fourth of these great inventions, the compass, was related with magnetism. The *Sinan* or *south-pointer* was made using lodestone, a special form of the mineral magnetite. Although it was the first compass, it was not a navigational instrument. Its first application was geomantic. The Chinese used it to order and harmonize their environments and buildings in accordance with the geometric principles of *feng shui*. Other types of compass, the *South-pointing fish* and *South-pointing turtle*, were used by jugglers as a form of entertainment. During the North Song dynasty, Shen Guo (1031-1095) conducted experiments with a magnetized needle that was used for navigational purposes in 12th century or even before.

It is necessary to distinguish between the compass and the mechanical *South-pointing carriage*, a differential gear mechanism invented by the Chinese between the 3rd and 2nd centuries BC.

Another notable China breakthrough highlighted on the syllabus is the *Technology of Iron and steel*. Chinese people very quickly developed the way to melt iron. They already had this technology in the 4th century BC, due to the following four reasons:

1. They had good refractory clays
2. They knew how to lower the temperature at which iron melts (from 1130 °C to 950 °C)
3. They had great experience with furnaces
4. They had the dual effect piston bellows

In this way they were able to obtain cast iron. They were also able to make many tools and even enormous objects like the Pagoda of Yu Quan (1061), all of them made in cast iron (13m). The first clear evidence of the fusion of wrought iron and cast iron to make steel comes from the 6th century AD. Apart from this co-fusion steel process, the Chinese had produced steel by the 2nd century BC through the process of decarburizing using the dual effect piston bellows to pump large amounts of oxygen into molten cast iron (the Chinese called this technique "the hundred refining method"). In part, this was possible because they had an air blowing device, the dual effect piston bellows (from 5th century BC), which ensured a continuous flow of air (Temple, 1986).

Another field is *the use of animal power*. Several Chinese inventions are introduced and emphasized here, such as the *stapes*, invented in the 3rd century (ancient armies, like the Egyptian, Babylonian, Greek or Roman did not have stapes ...); the *wheelbarrow* (1st BC- 1st AD), unknown in Europe until the 12th or 13th century, was a military secret and used in the Western Han dynasty. There were many different models of wheelbarrows in China; the *horse harness*, studied by Needham, who shows the different types: 1) the throat harness in the Ancient World, which greatly restricted the horse since the animal was in danger of choking; 2) the strap harness with a strap around the chest (1st century BC; Han dynasty), and 3) the horse collar, which enabled the horse to pull from the shoulders (4th to 1st centuries BC). It constituted a significant improvement in the use of animal power.

Achievements in the field of *navigation and nautical inventions* were astonishing. The Grand Canal, for instance, was as long as the distance between Rome and Paris, while the ships of the Ming Dynasty were enormous compared to ships of the same period in the West. Chinese ships of Ming fleet were technically far in advance of their European counterparts. (See the ICT Resources Annex 5.8 part 4/5 time interval: 5:15-8:57- & part 5/5 time interval: 0:00 – 1:00).

In the field of *Astronomy and astronomical instruments*, the early use of the equatorial coordinates placed Chinese astronomy in an advantageous position compared to European and Arabic astronomy (Ronan, 1993, vol.2). This made Chinese astronomical instruments very useful and easy to use. Chinese astronomers had at their disposal sophisticated observatories like the one at Gaocheng, designed by Guo Shoujing in 1276. In regard to *mechanical Clocks*, I show in the classroom the very sophisticated mechanism of the Su Song Clock (1092), which used an escapement mechanism for clockworks first developed during the Tang dynasty by Yi Xing, a highly skilful and ingenious Buddhist monk and court astronomer. (For more on Astronomy and Mechanical Clocks see the ICT Resources Annex, 5.8, part 2/5).

Needless to say, there is a wide field of Chinese inventions in the realm of *domestic technology*. In class, students study inventions such as Porcelain (3rd century), the umbrella (4th century), matches (6th century), etc.

In addition to these topics, students are also introduced to other Chinese technological achievements or inventions such as the Rotary Winnowing Fan (2nd century BC) for separating husks and stalks from the grain after it has been harvested; paper currency (8th-9th century); the open-spandrel segmental arch bridge, with one structure that is relatively lightweight thanks to semicircular arc spandrels, which enable high flood waters to pass through, such as those belonging to the Stone bridge over the river Jiao (from 610); or the Seismometer, invented by Zhang Heng (132 AD), the official astronomer and mathematician of the Eastern Han dynasty. This was an instrument which employed an inverted pendulum acting under inertia to release a metal ball by a trip device. This ball fell into the mouth of a corresponding metal object indicating the direction of tremors and earthquakes. Many other Chinese technological devices exist, some of which are as follows: the iron plow (4th century BC), the suspension bridge (1st century BC), waterwheels (1st century BC), the multiple groove planter (2nd century BC), the crank (2nd century BC), the fishing reel (4th century AD), the bridge suspended by iron cables (6th century), the transmission chain (979), the gimbal or "Cardan" suspension (9th century), etc.

As I mentioned above, as regards the use of online resources for teaching, the teacher usually chooses only a snippet of the video for the purpose of helping to understand a specific item on the agenda. For instance, if you are interested in the Zhang Heng seismometer or the South-pointing Chariot, then you should set the exact time interval on the video website. For these two examples, the start and end points are indicated below.

5.6 Eng [9:56] INVENTIONS OF THE GREAT ANCIENT CHINESE EMPIRE (3 OF 5)
http://www.youtube.com/watch?v=pJFQzGV85x0
Seismometer. Time interval 5:12-7:18

5.6 Eng [9:58] INVENTIONS OF THE GREAT ANCIENT CHINESE EMPIRE (4 OF 5)
http://www.youtube.com/watch?v=Ec9OD_uCueg
South-pointing Chariot. Time interval 5:40-6:16

Final considerations

To conclude this paper, I would like to recall the words of Chancellor Francis Bacon (1561-1626), a major supporter of the experimental method in 17th century:

> It is well to observe the force and virtue and consequence of discoveries, and these are to be seen nowhere more conspicuously than in **those three which were unknown to the ancients, and of which the origins, although recent, are obscure and inglorious; namely, printing, gunpowder, and the magnet.**

> For these three have changed the whole face and state of things throughout the world; the first in literature, the second in warfare, the third in navigation; whence have followed innumerable changes, insomuch that no empire, no sect, no star seems to have exerted greater power and influence in human affairs than these mechanical discoveries (Bacon 1620, 4, p. 114)

Although Bacon was unaware of the origins of these inventions, they are well-known today and are far from obscure; indeed, they were hardly "recent" in Bacon's time and originated in China.

The Chinese invented Technologies involving mechanics, hydraulics and mathematics applied to horology, metallurgy, astronomy, agriculture, engineering, navigation, and so on. In Ancient China, remarkable and useful technical discoveries were made for the benefit of all mankind.

A cultural gap exists in scientific and technical education where Western Science and Technology are concerned. Names such as Cai Lun, Bi Sheng, Wang Zhen, Ge Hong, Sun Simiao, Shen Guo and many others are quite unknown by European secondary-school students.

Teaching the History of Science and Technology of Ancient China can help students to broaden their technological knowledge as well as their culture. Additionally, it can facilitate the positive reception of other cultures, and therefore assist in the integration of foreign students

Fortunately, today's teachers have at their disposal a wealth of online resources. The ICT resources constitute a powerful tool that facilitates both teaching and student learning. All that is required for its implementation in teaching is a suitably equipped classroom, that is, one with an Internet connection and a good audiovisual system. You will find suggestions for Web addresses in the ICT Resources Annex. It may be useful in various parts of the curriculum.

References

Bacon, F. (1620), *The New Organon*. In Spedding, J., Ellis R., &. Heath D. (eds.), *The Works of Francis Bacon* (1887-1901), Mifflin and Company, Boston, vol. 4.

Chemla, K.& Shuchun, G. (eds.) (2005), Les Neuf Chapitres, le classique mathématique de la Chine ancienne et ses commentaires, Dunod, París.

Cheng, A. (1985), Histoire de la pensée chinoise, Éditions du Seuil, Paris.

Cullen, C. (1996), *Astronomy and Mathematics in Ancient China: The Zhou bi suan jing*, Cambridge University Press, Cambridge/New York.

Doroudi, S. (2007), *On Leibniz and the I Ching*. California Institute of Technology - Information Technology Services, 26 Apr. 2007

[on line: http://www.its.caltech.edu/~sdoroudi/LeibChi.pdf] (15/01/2011)

Fazzioli, E. (1987), Caractères chinois du dessin à l'idée, 214 clés pour comprendre la Chine, Flammarion, Paris.

Fairbank, J. K. (1996), *China, una nueva historia*, Andrés Bello, Barcelona-Buenos Aires.

Liu, G. & Zheng, R. (1989), *L'histoire du livre en Chine*, Éditions en Langues Étrangères, Beijing.

Martzloff, J.-C. (2006), *A history of chinese mathematics*, Springer-Verlag, Berlin Heildelberg [French version: (1987) *Histoire des mathématiques chinoises*, Masson, Paris].

Needham, J. (1954-2008), *Science and civilisation in China*, Cambridge University Press, Cambridge, 27 vols.

Puig-Pla, C. (2007), 'Ciència i Tècnica a l'Antiga Xina: Eines per al coneixement i l'aproximació cultural'. In Grapí P. & Massa M.R. (cords.), *Actes de la II Jornada sobre la història de la ciència I l'ensenyament Antoni Quintana Marí*, Societat Catalana d'Història de la Ciència I de la Tècnica, Barcelona.

Ronan, C. A. (1978–1995), *The Shorter Science and Civilization in China* [with Joseph Needham], Cambridge University Press, Cambridge, 6 vols.

Temple, R. (1986), The Genius of China: 3,000 years of Science, Discovery and Invention, Simon and Schuster, New York [Spanish version: (1987) El genio de China. Cuna de los grandes descubrimientos de la humanidad, Debate/Círculo, Madrid].

Zhuang, Wei (1980), *Cuatro grandes inventos en la antigüedad china*, Ediciones en Lenguas Extranjeras, Beijing.

ICT RESOURCES ANNEX

Below is a list of some ICT resources to help in teaching. They consist mainly of videos from YouTube. They are arranged thematically according to the syllabus of the subject mentioned, and contain references to the language used by each on-line resource, the total time of each resource in minutes (in square brackets), the title as it appears, the website address and a brief outline of the content.

Abbreviations: *Eng* = English; *Sp* = Spanish; *Ch* = Chinese; # = Background music without narration; * = No sound; *Eng Subt* = English subtitles; *Sp Subt* = Spanish subtitles; *Ch Subt* = Chinese subtitles. All Web addresses were active on 15 January 2011.

1. POPULATION AND TERRITORY

1.1 Eng/ Ch Subt [9:57] DISCOVERY CHANNEL, CHINA 1
http://www.youtube.com/watch?v=ErbMJAXr0Yw
Eng/ Ch Subt [10:03] DISCOVERY CHANNEL CHINA 2
http://www.youtube.com/watch?v=nCjW4ZxY2Bc
Eng/ Ch Subt [9:55] DISCOVERY CHANNEL CHINA 3
http://www.youtube.com/watch?v=gLuoIIYTan4
Eng/ Ch Subt (10:02) DISCOVERY CHANNEL CHINA 4
http://www.youtube.com/watch?v=-AQq2nrJAz8
Eng/ Ch Subt (10:05) DISCOVERY CHANNEL CHINA 5
http://www.youtube.com/watch?v=Yn01NDeK29k
Eng/ Ch Subt (10:01) DISCOVERY CHANNEL CHINA 6
http://www.youtube.com/watch?v=vZEYa8NXpy4
Eng/ Ch Subt (9:57) DISCOVERY CHANNEL CHINA 7
http://www.youtube.com/watch?v=4cCBLWeL14E
Eng/ Ch Subt (9:57) DISCOVERY CHANNEL CHINA 8
http://www.youtube.com/watch?v=gMR7LWzcYFw
"Set of 8 videos giving an overview of China today"
1.2 Sp. [1:54] DOCUMENTAL SOBRE CHINA. GEOGRAFIA
http://www.youtube.com/watch?v=47RD1N1-EU0
"Brief description of the territory of the Republic of China"
1.3 Sp [11:33] DOCUMENTAL GRANDES CIVILIZACIONES: CHINA
http://www.youtube.com/watch?v=g0N5pb2U2uo
"Territory, geography, resources and history up to the Qing Dynasty "
1.4 Sp. [1:51] EL HOMBRE MÁS ALTO DEL MUNDO VIVE EN CHINA Y AHORA ES PADRE
http://www.youtube.com/watch?v=XYx3I-4W-MY
"News about the world's tallets man, who lives in China"

299

1.5 # [8:48] WELCOME TO CHINA
http://www.youtube.com/watch?v=RowYdhAPEJU
"Video provided by the Chinese government during the World Exhibition Expo-Shanghai 2010 (culture, ethnicity, urban development, economics, etc.)"

1.6 # [3:11] GUILIN A TRUE BEAUTIFUL PLACE
http://www.youtube.com/watch?v=T4b6F9II17k
" "A beautiful set of static pictures of Guilin"

1.7 Eng [7: 23] MR. BRUSH'S CHINA GEOGRAPHY VIDEO
http://www.youtube.com/watch?v=KaZ9FP4mfu4
"A short overview of the geography of China"

1.8 Ch/Eng subt [1:59] CHINESE GEOGRAPHY
http://www.youtube.com/watch?v=x24fPPSYI-o
"About water resources in China"

1.9 Sp [1:36] LA GRAN MURALLA CHINA
http://www.youtube.com/watch?v=jxzgB9yktbg
"About the Great Wall"

1.10 Sp [0:46] LA GRAN MURALLA CHINA MIDE 8.851,8 KILÓMETROS 2.000 MÁS DE LO QUE SE ESTIMABA
http://www.youtube.com/watch?v=4WmNUAhi7sY
"On the length of the Great Wall"

1.11 Eng/ Ch-subt [9:02] [DISCOVERY]7 WONDERS OF CHINA 中国七大奇观 PART 1/5
http://www.youtube.com/watch?v=69pMsNaLSfk&feature=related

Eng/ Ch-subt [9:02] [DISCOVERY]7 WONDERS OF CHINA 中国七大奇观 PART 2/5
http://www.youtube.com/watch?v=j_V_Kc2-LFQ&feature=related

Eng/ Ch-subt [9:02] [DISCOVERY]7 WONDERS OF CHINA 中国七大奇观 PART 3/5
http://www.youtube.com/watch?v=0Tsk9nU-6s0&feature=related

Eng/ Ch-subt [9:02] [DISCOVERY]7 WONDERS OF CHINA 中国七大奇观 PART 4/5
http://www.youtube.com/watch?v=CrWG-KAaPI4&feature=related

Eng/ Ch-subt [8:59] [DISCOVERY]7 WONDERS OF CHINA 中国七大奇观 PART 5/5
http://www.youtube.com/watch?v=gB4j85XaKWM&feature=related
"About the 7 wonders of China: Terra Cotta Warriors (Xi'an, Shaanxi Province); Hanging Monastery (Mount Hengshan, Shanxi Province); The Great Wall Jiayuguan Pass (Gansu Province –West-) to Shanhaiguan Pass (Hebei Province –East-); Leshan Buddha (Leshan City, Sichuan Province); Mount Wudang, (Wudang, Hubei Province); Shi Bao Zhai Temple (Yangtze River -south bank-); Forbidden City (Beijing, China)"

2. CULTURAL FEATURES

MANDARIN LANGUAGE
2.1 Eng [5:07] LESSON 1 - INTRODUCTIONS - PART 3
http://www.youtube.com/watch?v=Ctvv12hVQAg
"A short lesson on Mandarin pronunciation (the four tones of Mandarin, the names of countries, etc.)"

2.2 Ch [3:47]MANDARIN LESSON #1
http://www.youtube.com/watch?v=XSiHnonlVjg

"About how to pronounce some Mandarin Pinyin Syllables"

2.3 Sp [7:233] COMUNICATE EN CHINO - LECCIÓN 1ª
http://www.youtube.com/watch?v=ELPVwPzHUNc
"A first lesson in Mandarin"

2.4 Ch [0:20] 七田真杜曼右腦閃卡注音符號
http://www.youtube.com/watch?v=JE-K2LShGPQ
"Bopomofo, the first official phonetic system for transcribing Chinese (sound and spelling)"

2.5 Ch [0:56] ㄅㄆㄇ之歌
http://www.youtube.com/watch?v=_Re32eyGwRo&feature=related
"Singing and spelling Bopomofo"

2.6 Ch – Sp subt [1:16] CURSO DE CHINO NUMEROS DE 1 A 10
http://www.youtube.com/watch?v=JzquEYUWNJM
"The pronunciation of the first 10 numbers in Mandarin and a final song!"

2.7 Ch – Sp subt [5:32] APRENDER CHINO: NUMEROS I - 数字 (ESCRITA)
http://www.youtube.com/watch?v=n00PoxztNpY
"How to write and pronounce numbers in Chinese"

2.8 Ch [0:53] COUNTING IN CHINESE 1-100
http://www.youtube.com/watch?v=R1csp3EtePs
"Children reading aloud the numbers 1 to 100"

2.9 Eng [2:55] LEARN CHINESE, COUNTING NUMBERS
http://www.youtube.com/watch?v=EDtWozaufs0&feature=related
"Pronunciation and spelling of numbers and tens (up to 100)"

2.10 Sp [1:26] CLASES DE CHINO MANDARIN
http://www.youtube.com/watch?v=LxbrOgXQq3M
"A simple and clear first lesson in Mandarin"

2.11 Sp [2:28] CLASES DE CHINO MANDARIN 1
http://www.youtube.com/watch?v=wJDS-xnDimw
"A simple and clear second lesson in Mandarin"

2.12 Sp [1:03] PALABRAS UTILES EN MANDARIN 01
http://www.youtube.com/watch?v=K9azdXmp4Pg
"Some basic words in Mandarin"

2.13 Eng [2:03] MANDARIN LESSON #2
http://www.youtube.com/watch?v=heSArGwNH0Y
"Regular expressions, written in Pinyin, but without indicating the tones"

CHINESE WRITING

2.14 # [0:39] CALIGRAFÍA CHINA - LOS NÚMEROS
http://www.youtube.com/watch?v=pT8TNj1JiQo
"Numbers written in Chinese"

2.15 # [10:18] IDEOGRAMAS CHINOS
http://www.youtube.com/watch?v=FgvBDtBk61Y
"Ideograms and calligraphy. Video art rather than a Training video"

2.16 Sp [0:37] CALIGRAFIA CHINA
http://www.youtube.com/watch?v=qjTQbx6Hrm8
"A person writing calligraphic characters on the street"

2.17 Eng [2:35] WRITING CHINESE CALLIGRAPHY WITH STYLE: COMPARING PRINTED FONTS & CHINESE CALLIGRAPHY
http://www.youtube.com/watch?v=48bVNPT2fLs
"How to write Chinese characters in proper calligraphy style (by Bo Feng, a Chinese/English translator and interpreter)"

2.18 Sp [2:39] APRENDER CHINO: MARCAS
http://www.youtube.com/watch?v=NjuS0o2eqeY
"Writing, pronunciation and Pinyin transcription of certain brands (Coca-Cola, Burger King, Starbucks, etc.)"

2.19 Sp [8:41]APRENDER CHINO: PAÍSES Y CONTINENTES: EUROPA
http://www.youtube.com/watch?v=ji5fYQqsK88
"The name of European countries in Chinese"

2.20 #/Sp Subt [3:17] ANIMACION DE CARACTERES CHINOS
http://www.youtube.com/watch?v=okqZla0oaOE
"Video entertainment, on Chinese characters and how to remember them. With the transcript in Pinyin"

2.21 Ch/ Eng Subt [1:31] LEARN CHINESE CHARACTERS
http://www.youtube.com/watch?v=dVAaESeTXPE&feature=related
"Spelling and pronunciation of some characters"

2.22 Ch & Eng [2:43] LEARN CHINESE CHARACTER(3)
http://www.youtube.com/watch?v=CSVXDDvLeEI&feature=related
"Writing with chalk on the blackboard"

2.23 Ch & Eng [2:56] LEARN CHINESE CHARACTER(4)
http://www.youtube.com/watch?v=QQw-8vwsNZI&feature=related
"Writing with chalk on the blackboard"

CHINESE MUSIC

2.24 # [2:46] CULTURA CHINA
http://www.youtube.com/watch?v=xY7yrAJpLSI
"Different pictures of China with a Chinese music background"

2.25 # [2:20] ÓPERA CHINA DE PEKÍN/SPANISH TEXT/CHINESE OPERA BEIJING
http://www.youtube.com/watch?v=2tycVQNtTA4
"A series of images on the Beijing Opera"

2.26 Ch [2:13] ANCIENT CHINESE MUSICAL INSTRUMENTS THREE IDENTICAL FEMALE
http://www.youtube.com/watch?v=Or0LjnbQ8G0
"Triplets playing the popular Chinese "guzhen" string instrument and three different beggars playing old single string instruments. They are very ancient.

2.27 # [10:52] CHINESE MUSIC - TRADITIONAL PIPA SOLO BY LIU FANG 霸王卸甲 劉芳琵琶
http://www.youtube.com/watch?v=-ZmAgFyVo48&feature=fvw
"China Traditional music for pipa (Chinese lute). Soloist: Liu Fang"

2.28 # [6:20] CHINESE MUSIC
http://www.youtube.com/watch?v=oqKZm5FJNAw&feature=related
"Chinese music and pictures from China"

2.29 Ch [9:33] THREE CLASSIC CHINESE SONGS
http://www.youtube.com/watch?v=_xmZ3fGhtng&feature=related
"The Sighs of Grief, the Wandering Songstress, and the Four Seasons 1982"

2.30 # [0:46] JJ IN CHINA 2007: ERHU
http://www.youtube.com/watch?v=2Z2jLsuGJmY&feature=related
"Erhu - Chinese Traditional Music"

2.31 # [5:20] CHINESE ERHU MUSIC: 月舞 DANCE TO THE MOON 二胡：于紅梅、琵琶：趙聰
http://www.youtube.com/watch?v=iMhtQAMGkWY&feature=related
"Two Chinese women playing"

2.32 # [2:41] TRADITIONAL CHINESE MUSIC
http://www.youtube.com/watch?v=7okLGfDULmI&feature=related
"Music played by the Tianjin University of Finance and Economics Orchestra"

2.33 # [6:10] CHINESE MUSIC IN THE GOLDEN HALL OF VIENNA (ERHU)
http://www.youtube.com/watch?v=uLFAu54G_Ys&feature=related
"Chinese music in the Golden Hall of Vienna (erhu), Vienna New Year Concert 1998"

CHINESE POETRY

2.34 Eng [13:32]CLASSICAL CHINESE POETRY - TANG LIBAI
http://www.youtube.com/watch?v=dW4D41sTjbQ
"About Poetry during the Tang dynasty"

2.35 Ch/Eng Subt [9:20] Li Bai李白 two poems: 静夜思 黄鹤楼送孟浩然之广陵
http://www.youtube.com/watch?v=id6ViI08UE4&feature=related
"Two poems by Li Bai; also a song"

3. HISTORY

3.1 Eng [9:37] ENGINEERING AN EMPIRE - CHINA 1/5
http://www.youtube.com/watch?v=wAvRPqMQRK0&feature=related
"The history of China. Food production, the Engineer Li Bing and his channel, the unification of the territory by Qin Shi HuangDi. With comments by Peter Weller, Joanna Waley-Cohen, John Major…"

Eng [9:17] ENGINEERING AN EMPIRE - CHINA 2/5
http://www.youtube.com/watch?v=0rCXy7yKVN8&feature=related
"About the Qin dynasty. The invention of Deep drilling, the Great Wall, The first emperor's tomb"

Eng [9:31] ENGINEERING AN EMPIRE - CHINA 3/5
http://www.youtube.com/watch?v=NfU9j9slYAY&feature=related
"The Terracotta Warriors, the end of the Qin Dynasty. The Han Dynasty: enhancing the Great Wall. The increase in population"

Eng [6:58] ENGINEERING AN EMPIRE - CHINA 4/5
http://www.youtube.com/watch?v=c202QvQHn10&feature=related
"The Sui dinasty, the Yang Di emperor and the Grand Canal. The fleet of Zheng He (Ming Dynasty)."

Eng [9:23] ENGINEERING AN EMPIRE - CHINA 5/5
http://www.youtube.com/watch?v=4LStpqTmILE&feature=related
"The fleet of Zheng He, the extension of the Great Wall and the end of the Ming dynasty"

ZHOU DYNASTY

3.2 Eng [5:00] [CHINESECIVILIZATION] ANCIENT CHINESE BRONZE WEAPONS / WEST ZHOU DYNASTY
http://www.youtube.com/watch?v=x-tIa4RgZx4&feature=related
"About Ancient Chinese Bronze Weapons from the West Zhou Dynasty"

QIN DYNASTY

3.3 Eng [9:44] THE FIRST EMPEROR: THE MAN WHO MADE CHINA (PART 1) http://www.youtube.com/watch?v=VUJr1Y2PNPI&feature=related

Eng [9:36] THE FIRST EMPEROR: THE MAN WHO MADE CHINA (PART 2)
http://www.youtube.com/watch?v=rYCM64KTRX8&playnext=1&list=PL01E5BF7DAF0C4ED8
Eng [9:33] THE FIRST EMPEROR: THE MAN WHO MADE CHINA (PART 3)
http://www.youtube.com/watch?v=GurXkZ4tnhs&feature=autoplay&list=PL01E5BF7DAF0C4ED8&index=5&playnext=2
Eng [9:29] THE FIRST EMPEROR: THE MAN WHO MADE CHINA (PART 4) http://www.youtube.com/watch?v=oGbgjrgaCJ8&feature=related
Eng [9:37]THE FIRST EMPEROR: THE MAN WHO MADE CHINA (PART 5) http://www.youtube.com/watch?v=3TNVVhv-zxo&feature=related
Eng [9:50] THE FIRST EMPEROR: THE MAN WHO MADE CHINA (PART 6) http://www.youtube.com/watch?v=FaNwNkhQWTs&feature=related
Eng [9:41]THE FIRST EMPEROR: THE MAN WHO MADE CHINA (PART 7) http://www.youtube.com/watch?v=ZdTDikvWRYc&feature=related
Eng [10:00]THE FIRST EMPEROR: THE MAN WHO MADE CHINA (PART 8) http://www.youtube.com/watch?v=7PfiPK8aASI&feature=related
Eng [10:00] THE FIRST EMPEROR: THE MAN WHO MADE CHINA (PART 9) http://www.youtube.com/watch?v=SQewLihz4Fg&feature=related
"About Qin Dynasty and the First Emperor"

HAN DYNASTY

3.4 Ch [9:53] Eng Subt【千古帝王】 汉武帝 EMPEROR WUDI OF HAN DYNASTY 1/4 http://www.youtube.com/watch?v=X9k6Rrxdcjg&feature=related
Ch [9:56] Eng Subt【千古帝王】 汉武帝 EMPEROR WUDI OF HAN DYNASTY 2/4 http://www.youtube.com/watch?v=pQOyOyDI9ss&feature=related
Ch [10:13] Eng Subt【千古帝王】 汉武帝 EMPEROR WUDI OF HAN DYNASTY 3/4 http://www.youtube.com/watch?v=0dVnBv5O0Jg&feature=related
Ch [10:11] Eng Subt【千古帝王】 汉武帝 EMPEROR WUDI OF HAN DYNASTY 4/4 http://www.youtube.com/watch?v=9_yg94ReJDM&feature=related
"About Han Dynasty and Wudi emperor"

SILK ROAD

3.5 Sp [1:16] RUTA DE LA SEDA http://www.youtube.com/watch?v=C_aG3asFiGE
"Brief idea about the Silk Road"
3.6 Eng [8:06] SILK ROAD
http://www.youtube.com/watch?v=ZTZWGJavQDw&feature=fvw
"About the Silk Road"

TANG DYNASTY

3.7 Eng [8:49] [NEW_FRONTIER] CHINA DURING THE TANG DYNASTY 1/3 CHINESE CIVILIZATION
http://www.youtube.com/watch?v=qclrSiGr_BU&feature=related
Eng [8:49] [NEW_FRONTIER] CHINA DURING THE SONG DYNASTY, PART 1, 2/3 http://www.youtube.com/watch?v=5kz8x5b5oAI
Eng [8:49] [NEW_FRONTIER] CHINA DURING THE TANG DYNASTY 3/3 CHINESE CIVILIZATION
http://www.youtube.com/watch?v=LGD-pcW2n1o&feature=related
Eng [8:46] [NEW_FRONTIER] TANG DYNASTY PART 2, 1/3
http://www.youtube.com/watch?v=XWv5GYdN8BE

Eng [8:49] [NEW_FRONTIER] TANG DYNASTY PART 2, 2/3
http://www.youtube.com/watch?v=FQBXVQyvR9U
Eng [8:49] [NEW_FRONTIER] TANG DYNASTY PART 2, 3/3
http://www.youtube.com/watch?v=D6Ui6KUR48E

SONG DYNASTY

3.8 Eng [8:11] [NEW_FRONTIER] CHINA DURING THE SONG DYNASTY, PART 1, 1/3 http://www.youtube.com/watch?v=GpqUsyMeds4&feature=related
Eng [8:13] [NEW_FRONTIER] CHINA DURING THE SONG DYNASTY, PART 1, 2/3 http://www.youtube.com/watch?v=5kz8x5b5oAI
Eng [8:14] [NEW_FRONTIER] CHINA DURING THE SONG DYNASTY, PART 1, 3/3 http://www.youtube.com/watch?v=Dx--FKlEzm4
Eng [8:46] [NEW_FRONTIER] THE SONG DYNASTY, PART 2, 1/3
http://www.youtube.com/watch?v=HKXYB5FzbNU&feature=related
Eng [8:51] [NEW_FRONTIER] THE SONG DYNASTY, PART 2, 2/3
http://www.youtube.com/watch?v=mBRRxKglLD8&feature=related
Eng [8:56][NEW_FRONTIER] THE SONG DYNASTY, PART 2, 3/3
http://www.youtube.com/watch?v=4oHRGLPwbOg&feature=related
"6 videos about the Song Dynasty"

YUAN DYNASTY

3.9 Eng [10:00] DYNASTIES: THE MONGOL EMPIRE- PART 1/5
http://www.youtube.com/watch?v=znKNqj0-jGg&feature=related
Eng [10:00] DYNASTIES: THE MONGOL EMPIRE- PART 2/5
http://www.youtube.com/watch?v=zpvAvBwUXps&feature=related
Eng [10:00] DYNASTIES: THE MONGOL EMPIRE- PART 3/5
http://www.youtube.com/watch?v=A7mSxiO-Fvw&feature=related
Eng [10:00] DYNASTIES: THE MONGOL EMPIRE- PART 4/5
http://www.youtube.com/watch?v=WaJMIihcaEk&playnext=1&list=PL14B2D2E4811D32E1
Eng [8:03] DYNASTIES: THE MONGOL EMPIRE- PART 5/5
http://www.youtube.com/watch?v=Frp7bT0IJWk&feature=related
"5 videos about Kublai Khan and the Yuan Dynasty"

MING DYNASTY

3.10 Eng [8:26] [NEW_FRONTIER] THE MING DYNASTY, PART 1, 1/3
http://www.youtube.com/watch?v=l881NVLbrvY&feature=related
Eng [8:29][NEW_FRONTIER] THE MING DYNASTY, PART 1, 2/3
http://www.youtube.com/watch?v=81ZQvqnf76A&feature=related
Eng [8:28] [NEW_FRONTIER] THE MING DYNASTY, PART 1, 3/3
http://www.youtube.com/watch?v=nCAtvQQYd1U&feature=related
"3 videos about the Ming Dynasty"
3.11 Eng [5:52] ZHENG HE[1]
http://www.youtube.com/watch?v=pzFq0Ivwz9g&feature=related
"About Zheng He"

---

[1] About Zheng He, see also in History: Eng [9:23] ENGINEERING AN EMPIRE - CHINA 5/5.

305

QING DYNASTY
3.12 Eng [5:34] THE QING DYNASTY PART 1
http://www.youtube.com/watch?v=FPovbqhG1TA&feature=related
Eng [6:02] THE QING DYNASTY PART 2
http://www.youtube.com/watch?annotation_id=annotation_95574&v=5kUJtNRtIkI&feature=iv
"About the Qing Dynasty"
3.13 Sp [1:35] CHINA EN EL SIGLO XVIII
http://www.youtube.com/watch?v=8w-GEX2wYNE
"About the Qing Dynasty (17th and 18th centuries)"

4. TRADITIONAL CHINESE THOUGHT

4.1 Eng [8:46] 1 OF 5 CONFUCIUS BIOGRAPHY
http://www.youtube.com/watch?v=NCqIjq6ff-k&feature=related
Eng [8:07] 2 OF 5 CONFUCIUS BIOGRAPHY
http://www.youtube.com/watch?v=OgcFscScRK0&feature=related
Eng [9:04] 3 of 5 Confucius Biography
http://www.youtube.com/watch?v=XWQ2vwZxiD4&feature=related
Eng [9:05] 4 OF 5 CONFUCIUS BIOGRAPHY
http://www.youtube.com/watch?v=68Npq2L0GtQ&feature=related
Eng [9:14] 5 OF 5 CONFUCIUS BIOGRAPHY
http://www.youtube.com/watch?v=TtLdgNYymdE&feature=related
"On Confucius' biography"
4.2 Eng [4:31] CONFUCIANISM AND ITS IMPACT ON CHINA
http://www.youtube.com/watch?v=QjTVSNtFzUg&feature=related
Eng [4:15] 5 RELATIONSHIPS OF CONFUCIANISM
http://www.youtube.com/watch?v=Nn7grjTpcNA&feature=related
Eng [0:59] 5 RELATIONSHIPS OF CONFUCIANISM - THE ANSWER
http://www.youtube.com/watch?v=AhkdxIVln7Y&feature=mfu_in_order&list=UL
Eng [3:43] INTRO INTO CHINESE LECTURE
http://www.youtube.com/watch?v=Kqs268zhQd4&feature=BF&list=ULvK3J-KF0nxE&index=3
4.3 Eng [1:26] INTRO TO DAOISM
http://www.youtube.com/watch?v=Oi62EjwG__8&feature=autoplay&list=ULvK3J-KF0nxE&index=2&playnext=1
"Intro to Daoism. The symbol of Yin Yang"
4.4 Eng [6:06] TAO TE CHING (PT.1)
http://www.youtube.com/watch?v=1BSiZQqlg5E&feature=related
Eng [6:20] THE TAO TE CHING (PT. 2)
http://www.youtube.com/watch?v=lPWY-fFCbRc&feature=channel
Eng [6:50] THE TAO TE CHING (PT. 3)
http://www.youtube.com/watch?v=hyzAmP3B8vU&feature=channel
"A reading of the text of the Tao Te Ching"
4.5 Eng [6:00] TAOISM - PART 1
http://www.youtube.com/watch?v=lynaDSQ0V0Y&playnext=1&list=PLE4C35FA8788E2FEC
Eng [6:00] TAOISM - PART 2
http://www.youtube.com/watch?v=bjLVr1fG5Ec&feature=related
"2 videos about Taoism"

4.6 Eng [5:00] [CHINESECIVILIZATION] LAOZI AND DAO DE JING / 老子和道德经
http://www.youtube.com/watch?v=bMcTL8qdik0&feature=related
"About Laozi and Dao De Jing"
4.7 Eng [5:16] LAOZI AND DAODEJING
http://www.youtube.com/watch?v=zsNczpM5pao&NR=1
"About Laozi and Dao De Jing"
4.8 Sp [1:38] DIFUSIÓN DEL BUDISMO
http://www.youtube.com/watch?v=y1Q0E_m_xZQ
"The spread of Buddhism in Asia"
4.9 Eng [8:38] HISTORY SERIES - BUDDHISM, ORIGINS AND INTERACTION WITH GREEK PHILOSOPHY
http://www.youtube.com/watch?v=3eH8sawewws&NR=1
"On the origins of Buddhism"
4.10 Eng [10:09] HISTORY SERIES - BUDDHISM, VIRTUES, TYPES AND CENTER OF LEARNING
http://www.youtube.com/watch?v=IOlBsLAZyoc
"About Buddhism: virtues, types, learnig"
4.11 Eng [3:56] TYPES OF BUDDHISM
http://www.youtube.com/watch?v=wsICPwfiYlQ&feature=related
"Different types of Buddhism"
4.12 Sp [8:45] HISTORIA DEL BUDISMO 01
http://www.youtube.com/watch?v=Pyi_jA-wZ24
Sp [8:52] HISTORIA DEL BUDISMO 02
http://www.youtube.com/watch?v=OGlnrnzuwUY
Sp [6:04] HISTORIA DEL BUDISMO 03
http://www.youtube.com/watch?v=PHVFfECOBk0
"3 Videos about Buddhist History"
4.13 Sp [9:53] EL BUDISMO ZEN - PARTE 1/2
http://www.youtube.com/watch?v=IIcnHbb20Dw&feature=related
Sp [9:53] EL BUDISMO ZEN - PARTE 2/2
http://www.youtube.com/watch?v=avLxxODPKEY&feature=related
"Buddhism explained through a conversation between two people"

5. INVENTIONS FROM ANCIENT CHINA

5.1 Eng [11:52] INVENTIONS FROM ANCIENT CHINA ~ COMPASS
http://www.youtube.com/watch?v=esFBpzwKLHM&NR=1&feature=fvwp
"The invention of the compass"
5.2 Eng [12:52] INVENTIONS FROM ANCIENT CHINA ~ PAPER MAKING TECHNIQUE
http://www.youtube.com/watch?v=uXNFq1bUoT8&feature=fvwrel
"About Chinese Paper Making Techniques"
5.3 Eng [9:25] INVENTIONS FROM ANCIENT CHINA ~ GUNPOWDER
http://www.youtube.com/watch?v=D6j5vlWNQc8&feature=channel
"The invention of Gunpowder"
5.4 Eng [16:04] INVENTIONS FROM ANCIENT CHINA ~ PRINTING TECHNIQUE
http://www.youtube.com/watch?v=Zw3Dn3eaSOI&feature=relmfu
"The invention of Chinese Printing Techniques"
5.5 Eng [10:08] HAN DYNASTY SEISMOGRAPH IN 132AD

http://www.youtube.com/watch?v=GcVFuIccf5c
"On the invention of the seismograph

5.6 Eng [(9:56] INVENTIONS OF THE GREAT ANCIENT CHINESE EMPIRE (1 OF 5) http://www.youtube.com/watch?v=lxceQ9-6uHQ
Eng [9:59] INVENTIONS OF THE GREAT ANCIENT CHINESE EMPIRE (2 OF 5) http://www.youtube.com/watch?v=jGAZgGAMlMY
Eng [9:56] INVENTIONS OF THE GREAT ANCIENT CHINESE EMPIRE (3 OF 5) http://www.youtube.com/watch?v=pJFQzGV85x0
Eng [9:58] INVENTIONS OF THE GREAT ANCIENT CHINESE EMPIRE (4 OF 5) http://www.youtube.com/watch?v=Ec9OD_uCueg
Eng [7:00] INVENTIONS OF THE GREAT ANCIENT CHINESE EMPIRE (5 OF 5) http://www.youtube.com/watch?v=ToFTgYUNCpc
"Very interesting set of 5 videos on Inventions of the Great Ancient Chinese Empire"

5.7 Eng [10:37] ANCIENT DISCOVERIES - MACHINES III 1/4
http://www.youtube.com/watch?v=A5yZbunOQ04&feature=related
Eng [10:37] ANCIENT DISCOVERIES - MACHINES III 2/4
http://www.youtube.com/watch?v=kt_nZoNbXrY&feature=related
Eng [10:37] ANCIENT DISCOVERIES - MACHINES III 3/4
http://www.youtube.com/watch?v=IgxCRlNuHpE
Eng [10:37] ANCIENT DISCOVERIES - MACHINES III 4/4
http://www.youtube.com/watch?v=xM68rdP-56g&feature=related
"About ancient Chinese Inventions and Technology"

5.8 Eng/ Ch-subt [8:57] [DISCOVERY]ANCIENT CHINESE TECHNOLOGY 中国古人的智慧 PART 1/5
http://www.youtube.com/watch?v=GTFeFKn_j-Q&feature=related
"Ancient Chinese Technology: Fire and Gunpowder"

Eng/ Ch-subt [8:57] [DISCOVERY]ANCIENT CHINESE TECHNOLOGY 中国古人的智慧 PART 2/5
http://www.youtube.com/watch?v=fRmjxl8zUIs&feature=related
"Ancient Chinese Technology: Water and Astronomy instruments. The mechanical clock"

Eng/ Ch-subt [8:57] [DISCOVERY]ANCIENT CHINESE TECHNOLOGY 中国古人的智慧 PART 3/5
http://www.youtube.com/watch?v=bGmoOX9ZGZE&feature=related
"Ancient Chinese Technology: Paper, woodcut, cast iron and the plow"

Eng/ Ch-subt [8:55] [DISCOVERY]ANCIENT CHINESE TECHNOLOGY 中国古人的智慧 PART 4/5
http://www.youtube.com/watch?v=n3ed49S-KzE
"Ancient Chinese Technology: Steel, the double-acting piston bellow, the crossbow, advances in navigation, the magnetic compass"

Eng/ Ch-subt [8:53] [DISCOVERY]ANCIENT CHINESE TECHNOLOGY 中国古人的智慧 PART 5/5
http://www.youtube.com/watch?v=PBY1BbbfyHw
"Ancient Chinese Technology: Earthquakes, the seismograph"

5.9 Eng [13:43] ANCIENT CHINESE ARCHITECTURE
http://www.youtube.com/watch?v=AgBA1veRlH0&feature=related
"About Chinese Architecture (Great Wall, buildings, bridges, etc.)"

5.10 Eng [9:12] Ch Subt CHINA'S MEGA DAM - PARTE 1 DE 9
http://www.youtube.com/watch?v=PPp248CoKLU
Eng [9:13] Ch Subt CHINA'S MEGA DAM - PARTE 2 DE 9

http://www.youtube.com/watch?v=o3dF6zHOybg&NR=1
Eng [9:13] Ch Subt CHINA'S MEGA DAM - PARTE 3 DE 9
http://www.youtube.com/watch?v=xy3miNRHSS8&feature=related
Eng [9:33] Ch Subt CHINA'S MEGA DAM - PARTE 4 DE 9
http://www.youtube.com/watch?v=aidTzgKXZ04&feature=related
Eng [9:33] Ch Subt CHINA'S MEGA DAM - PARTE 5 DE 9
http://www.youtube.com/watch?v=LEgFFTVEzFk&feature=related
Eng [9:33] Ch Subt CHINA'S MEGA DAM - PARTE 6 DE 9
http://www.youtube.com/watch?v=vy6GXnpV3_I&feature=related
Eng [8:28] Ch Subt CHINA'S MEGA DAM - PARTE 7 DE 9
http://www.youtube.com/watch?v=nFYUqWNipmw
Eng [8:29] Ch Subt CHINA'S MEGA DAM - PARTE 8 DE 9
http://www.youtube.com/watch?v=gsFsLasZlN0
Eng [8:28] Ch Subt CHINA'S MEGA DAM - PARTE 9 DE 9
http://www.youtube.com/watch?v=CjoJiDU56a8&feature=related
"About floods and the Three Gorges Dam"

The application of the history of science and the use of the Internet as methodological tool in the subject of sciences of the contemporary world

Raimon Sucarrats Riera & Agustí Camós Cabecerán

Centre d'Història de la Ciència (CEHIC), Universitat Autònoma de Barcelona.

ABSTRACT: The appearance in the Spanish educational system of a new subject, *Science for the contemporary world* (SCW), has opened the possibility to use the history of science and technology as a pedagogic tool to understand the nature of science and its challenges in the beginnings of the 21st century. This fact is happening in a world which is undergoing a change of paradigm in the transmission of information, caused by the internet in the knowledge society. After studying the curriculum of SCW, this paper exposes the methodological experiences of two teachers who try to join the science teaching of the present time using a historical vision with the resources that the internet provides to the educative community.

Teaching science in the secondary school

It must be taken into account that a relevant number of students don't study scientific subjects in their educational period before university. Consequently, they don't receive any information about cultural aspects related to science which would be of great significance in their daily life. This fact is a paradox, since science is a kind of knowledge that is believed with high credibility and it is considered to be of great importance for the future of any country. That is the reason why some European countries have made an effort to include in their educational curricula a subject with science contents in these pre-university years of students' formation.

In the case of the United Kingdom (Millar & Hunt: 2006), *Divulgative science* is a subject that was designed in 1996 in order to give enough scientific information to students older than 16 years old. So they would become more intelligent consumers of science and technology. Therefore, its programme includes the most popular scientific topics. That's to say those topics which appear more frequently in the mass media. There have been made different studies which have demonstrated that health and environment are the main ones, followed by space exploration and astronomy. However, learning scientific contents is not the only objective of this subject. It is also very important to show the students the real nature of science and the characteristics of scientific reasoning and methodology. For this reason, discussion of scientific topics among students must be fostered. They should also be taught to interpret scientific information taken out of newspapers, magazines, radio programmes, etc. The final goal is to educate people to be able to understand any article about divulgative science.

In France (Mas:2006), from 2000 onwards, students who do not choose scientific options in the *Lycée* (16-18 years old) have a multidisciplinary scientific learning with the objective to give them the necessary keys to understand the presence of science in our society. Here we can find again the need to educate citizens with some basic knowledge about science and its methods. In this way, these people will be able to comprehend and be critical about some problems of our society and about the responsibility of human's actions in the degradation of our World.

In Spain, the *Ley Orgánica de Educación* (Organic Law of Education) in 2006 anticipated the existence of a new subject, *Science for the contemporary world* (SCW). This subject attempted, in some way, to give this scientific basis we have talked before to the students of baccalaureate (16-18 years old). Since this new subject was proposed, many experts have tried to define the main aspects of its curriculum (Pedrinaci: 2006). From the very beginning there were doubts about which contents should students receive in order to make their integration in our scientific and technologic society easier. On the other hand, it has been discussed how to give coherence to SCW curriculum. Everybody agreed that it would be not enough to teach a collection of different scientific topics which could be interesting in our time. On the contrary, it would be better to educate citizens who were able to take intelligent decisions and to understand how science thinks and works. In our opinion, the use of history of science and technology contributes to build a solid basis to let knowledge grow regarding what science is and represents[1].

This historical method can be used as introduction to any scientific topic. It can also indicate where its knowledge comes from, how it evolves, its importance and interdependence with other cultural facts, as well as establishing which intellectual and methodological challenges must face up to (Izquierdo et al.: 2006).

Even though some authors are optimistic about the use of history, and even philosophy, in science classes, we consider that it is quite probable that this use is merely anecdotic. Teachers use normally this resource only to describe the *heroes or fathers* of the discipline and they commonly interpret the events of the past from the reality of the present, as if there was a predetermined direction in history. This way to look at the past is very well known as *whiggism*. That is the reason why a good training of educators in this field becomes so important. They should have enough knowledge in historiography to promote a correct use of the history of science. Then, they would avoid teaching history in a *whiggist* way, without an appropriate contextualisation of facts. For this reason, many universities have developed degrees of education in the history of science. In the Catalan case, the Centre d'Estudis d'Història de la Ciència (CEHIC, History of Science Studies Centre) in the Universitat Autònoma de Barcelona (UAB, Autonomous University of Barcelona) has developed a Master in History of Science and Technology. In this master, a limited number of teachers have obtained a higher degree in this topic. Moreover, there have been organised more popular courses addressed to teachers in secondary levels. For example, the course *Science and Technology through History* which started in 2009 and has been followed by a significant high number of teachers who are interested in the application of history in their classes of mathematics, chemistry, physics, biology or geology (See Pere Grapi's article in this same work). This course has been designed online, using the MOODLE (Module Object-Oriented Dynamic Learning Environment) as an educational platform. However, unless the history of science had a consolidated position in university degrees, normalization of its use in the scientific education of secondary students would not be possible. Unfortunately, it is not the case.

Syllabus of *Science for the contemporary world* and expectations for the history of science and technology

On July 29th 2009 the Government of Catalonia published the *Decree of ordination of higher education*. In this decree it was described the new subject of *Science for the contemporary world*. This subject should be given to all students that coursed the first level of the baccalaureate (16-17 years old), two hours per week all along the official course. The decree emphasizes the fact that this subject appears just in the moment in which the new paradigm of knowledge and information society is being imposed. The most remarkable proposal of the subject is related to the "acknowledgement of the universal character, not compartmentalized, of science as a product of culture, as a collective work of thinking and imagination which free us from ignorance and superstition, and as an activity that makes the efficient satisfaction of human necessities possible through its technological application". There are other important objectives of this subject, such as teaching the way a scientific works or showing the narrow relationship between scientific activity and quotidian interests of population. It can be taken into consideration that this subject is going to be explained to all kinds of students, either those who have chosen a scientific option or those who have chosen a humanistic one. So, the contents and activities must be adapted to that diversity and must not be exhaustive.

The syllabus of SCW is divided into five sections: origin and evolution of Universe; science, health and styles of life; human development and sustainable development; materials, objects and technologies; and information and knowledge. These sub-headings have the only function to make more comprehensible the subject to let the teacher explain the knowledge and methodologies of the science of our time. Let's see a summary of the contents that should be discussed in every one of these sections:

The section *Origin and evolution of Universe* discusses topics like the Big Bang theory, plate tectonics theory, and the origin and evolution of living beings, especially the human species. The section *Health and styles of life* is suitable to talk about topics related to metabolism, genetics, biotechnology, health and sickness, medical sciences, etc. In the one which is entitled *Human development and sustainable development* we will find the right place to explain the interaction between humans and the environment, and to discuss topics like global warming or the lost of biodiversity. In the section *Materials, objects and technologies* there can be analysed physical and chemical properties of materials and their relationship with their utility, and there can also be shown new technologies such as nanotechnology or taken into account the life cycle of materials and waste treatment. And finally, the section *Information and knowledge* should emphasize the new social and cultural paradigm based on the application of information and communication technologies (ICT).

Nevertheless, the most interesting aspect of this subject is not found in its contents, which as it has been said should not be treated in an exhaustive way, but in the basis of what we consider that should be transmitted to our students. The question is not to teach only concepts and methods, but to try to explain the real nature of science, its transcendence in our daily life and its ethical implications. That is the reason why this decree specifies that this subject has to improve students' abilities in searching information, reflecting on the nature of science and understanding the social and civic dimension of science and technology. Furthermore, the decree recommends the implementation of didactic activities in order to promote the use of sources of information, develop limited researches, work in little groups, and potentiate debates and presentations to classmates. This didactic philosophy should also affect evaluation activities. These activities should combine the traditional exams with other methods which could foment intellectual creativity, cooperation and solidarity.

With all this record, it is not strange that SCW can be linked with other subjects. And not only can be linked with scientific or technologic subjects, but also with humanistic ones such as history, philosophy or citizenship. One of the objectives of the subject claims: "Get to know general premises of the main cosmologies and unifying scientific theories, the cultural and historical context in which they were formed and the controversies they caused." This is the cause why teachers with some interest in the history of science and technology consider SCW an excellent frame to explain the present state of science as a result of its history. However, an accurate reading of its contents may disappoint those teachers. We don't find in its curriculum many topics where we could apply this historical vision. In fact, there can only be found few points with this type of references:

- Discuss the change of paradigm in relation to the representation of Universe: from geocentricity to heliocentricity.
- Debate the theories on the origin of life.
- Consider the epidemic phenomenon in the historical context and in our time.
- Characterize the environmental crisis throughout history.
- Humankind and the use of materials

Anyway, this list can be completed with many other topics which can also be explained from a historical point of view: Big Bang theory, plate tectonics theory, medicine and pharmacy history, and so forth. Moreover, there is a common topic that is shared by all the others. It is indicated in the objectives of the subject, but does not appear explicitly in the curriculum. This is the nature of science: the characteristics of its knowledge, its methodology and its social and ethical implications. This topic requires, without any doubt, the use of examples coming from the history and the philosophy of science.

Once the SCW's curriculum was known, even before, the publishers specialized in textbooks began the task of designing appropriate materials to help teachers to teach this subject. The final result seems demonstrate that it has not been a general consensus about which contents are needed to be included in an ideal SCW's textbook. There is a great diversity among SCW's textbooks that have been published. We don't know the cause of it. It could be either the lack of time or the fact that there is no tradition in this type of subject. There are other traditional subjects, for example biology, where we can detect a common discursive line in the great majority of textbooks, regardless of the publisher. On the contrary, the contents and activities proposed for SCW only coincide in the big titles, imposed by the official curriculum. In any other aspect, we can find a huge diversity.

We were especially interested in one characteristic of these books. That was obviously how they used the history of science and technology as an essential tool for the explanation of their different contents. In order to make a significant study, eight SCW's textbooks have been revised[2]. Although it was not our intention to make an exhaustive comparative among them, we have found few uses of the history of science and technology. Let's see some data:

- Only three of these eight books have an introductory chapter dedicated to the nature of science and technology and their social implications.
- None of them uses the history as a thread of the subject. There is only one that gives some importance to history in a significant number of chapters.
- Normally, the few references to the history of science and technology, when they are present, are located in a secondary space of the book or in one activity.

Besides, the historical topics that are used in these textbooks are the typical ones, those that are found in the textbooks of other subjects such as biology or geology. The two highlights, those topics that are always explained from their history, are the origin of life (spontaneous generation, the experiments of Urey and Miller, etc.) and evolution (creationism, Lamarckism, Darwinism, synthetic theory). There are a couple of topics whose history is explained by four of five of the books. These are plate tectonics (great geological models, continental drift) and the origins of genetics, with his *father* Mendel. And that's all. We can scarcely find other historical references in one or two of them. It's rather surprising that only one book uses directly the debate between geocentric and heliocentric theories to introduce the Big Bang and other modern theories about the origin of Universe. To give other examples, only one textbook talks about the history of medicine and pharmacy, and such important personalities for the history of science as Aristotle or the Count of Buffon can hardly be found in the pages of these textbooks.

It is obvious that if we want to use history as an important educational strategy in SCW, we have to search other sources of information. Being a fact that knowledge society and information and communication technologies constitute one of the five sections of SCW's curriculum, it is a good opportunity to use the internet as a decisive support of our educational task. The same way as it is important in so many activities in the society of our time. The use of the internet has not to be reduced to the simple transmission of contents. It must include the possibility of interaction by the students and has to collaborate decisively in their evaluation. The workshop *Mind the gap* that has been held in the University of Brest on March 2010 has proved that there is a great amount of resources in history of science and technology which can be found in the internet. Now we will study the application of this type of methodologies in two different cases.

First experience: History as the backbone of SCW teaching

The two next sections will explain the teaching experience in SCW that is being applied by one of the authors of this paper, Raimon Sucarrats, since 2008-2009. Just from the beginning, we thought that SCW opened the possibility for all students to achieve a minimum of scientific culture to let them appreciate the nature of scientific work. As we have said before, students who are just about to start their studies in the university must be able to understand a divulgative text on science. It should be equally assured that these students know the most popular scientific theories, such as plate tectonics or evolution. Therefore, this opportunity given by the new curricula to teach scientific culture to these students should not be wasted. Moreover, we believed that we could make the most if the history of science was used in SCW as a basis of our explanations and ICT as the main system of transmission of information. Our idea was to face the subject with a determined methodology: always taking a significant historic event, conveniently contextualized, as starting point of explanation, and emphasizing the use of ICT and the internet as an essential way of consultation and learning.

Once these methodological principles were established, the first decision to take in order to design the classes in the academic year 2008-2009 was to establish whether it was necessary a textbook or not. We have already explained the characteristics of the textbooks that have been published and their lack of historical contents. Following the pedagogic criteria we had proposed, none of them included the contents and activities we wanted to apply. As a consequence of that, we decided not to use a textbook and so we started searching for information on the internet in order to substitute the function of the textbook with the pedagogic materials we could find there. Once this decision had been taken, it was decided that the teacher would teach using these materials found on the internet with the help of a projector. Students could consult those contents at home after the class. Moreover, other uses of the internet were strengthened. First of all, a forum of the subject was created in the intranet of the school. Then, as we will analyze later, a blog was designed to let students find more information, participate in debates and freely give their opinions. It was also taken into account that evaluation criteria should agree with the rest of methodology and it had not to be exclusively based on traditional written exams. Students' participations in online activities and other type of works ought to have a real transcendence in the final qualification. We should assure that this evaluation system promoted student's creativity and prepared them for future challenges in the university. It was also necessary to promote autoevaluation and also let students criticize the whole methodology. The academic year 2008-2009 started with these premises.

Through the two first courses there have been changes in the methodology as a consequence of the experience of the teacher and the reception and analysis of constructive criticisms from students. Now we will explain the methodology that is been applied this year 2010-2011 as a result of this process of changes:

A complete chapter is dedicated to explain the characteristics of science, which are its objectives and methodology, and to reflect on the ethical repercussions of its activity. The same process is done with technology. In this chapter we use the history of science as a pedagogic resource. For example, it's explained the case of Dr. Semmelweis[3] in the 19th century in Vienna. His hazardous life, the methodology he developed in order to infer his conclusions and the reception of his results by the community of doctors are wonderful tools to explain how science works and evolves. This case also shows the importance of the context in scientific activity: paradigmatic knowledge difficult to eradicate, dogmatism, scientific rivalry, scientific nationalism, etc.

Every section of the programming of the subject is always started by a historical reflection. A brief summary of the history of science or a significant case can be explained to make students understand how science worked in that period and, consequently, how it works today. At the same time, students understand the real nature of science while listening to contextualized histories about the evolution of scientific activity in the past. We should try to avoid visiting the past as it was an incomplete project of the present. The goal is to immerse ourselves in a different world where scientific activity is one more element of the cultural network of that time. Let's see some examples which are explained in this subject that are not very common in textbooks:

- The conception of our time about the way the Universe appeared is introduced from the cosmology of Aristotle. Later, there are explained the systems of Ptolemy and Copernicus. To finish with, there is a debate about the Big Bang theory and the proofs that seem demonstrate it.
- The idea of evolution let us revise the extraordinary lives of such great personalities as Lamarck and Darwin. We must wonder why those things happened in that specific century and which the causes of the crisis of creationism were. Moreover, we can revise which circumstances lead to the existence of the theory of intelligent design.
- The section of health and technology deals with genes, heritage, diseases, etc. We start this section with an explanation about the idea of *the chain of beings* and explaining in which different manners philosophers understood the idea of species in 18th century. It can be taken as an example the polemics between Carl von Linnaeus and the Count of Buffon. Here we should discuss about whether species exist or not, and the significance of taxonomy and the nomenclature of species. On the other hand, a brief history of medicine and surgery is quite enlightening of the evolution of the medical profession from the Greek or Roman periods to nowadays. For example, the wonderful pages of *De humani corporis fabrica* by Vesalius wake students up to a great interest in the change of anatomy in 16th century, and remit to personalities of the antiquity as Hippocrates (it's very interesting to show the Hippocratic oath) or Galen.
- Since environmental problems of our time have been discussed in other subjects of the curricula, it's interesting to explain how environment has changed in the past when ecological conscience was not very extended in western societies.
- The section dealing with technology begins with a brief vision of the evolution of materials and the way these materials have been used in different periods of humankind.

We have already said that students don't have a textbook to consult. So, they have to take notes of the explanations of the teacher. However, this explanation is usually completed by contents which can be found on the internet. It's known that teachers can select a lot of didactical materials from the internet. They can use specialized webs that bring high level information of whatsoever field we could imagine or even fragments of films (YouTube) that can easily contribute to create debates among students if they are finely selected. We can also find books, biographies, dictionaries, webs of museums, etc.

In our case, we usually start every chapter with the projection of a fragment of a documentary or a film in order to focus the main part of the topic. The use of fragments of commercial films is quite interesting to show a set of quality movies that deal in a certain way with scientific topics to the students. In every case, students know the address on the internet of all these pedagogic materials so that they can consult them at home whenever they want.

Using a blog as an educational tool

The use of a blog specifically designed as an educational tool has been the last innovation in our way to teach SCW (See fig. 1). The main idea was to offer quality information to pupils in a different format, nearer and easily available for them, and to give them the possibility of participating actively in the subject. In fact, as we will see later, a significant part of the evaluation of the subject is based on students' participation in the blog. Let's see how it works[4]:

The blog contains three pages where students can find different texts. They must read them and, at least, they have to comment one of them. These texts are monthly renewed. In this way, every student has to comment, approximately, two or three of these texts every evaluation term. Texts are taken from classic books, scientific papers, scientific news in newspapers, textbooks, etc. Whatever resource it could be, the text must have an interest for students and an adequate level for them. These are some examples of texts that have been selected until now. Obviously, all of them are fragments of the complete work:

- An essay of Michael de Montaigne's *Essais* where he deals with the rightness of scientific opinions related to the new ideas, in that time, of Copernicus.
- The dramatic play *Galileo Galilei* by Bertolt Brecht. Just the moment in which the Italian saviour retracts of his theory.
- The famous paper in *Nature* (1953) by James Watson and Francis Crick, where they describe the double helix structure of DNA.
- Historic books by Buffon (*Histoire naturelle*), Lamarck (*Philosophie zoologique*), or Darwin (*Autobiography* or *On the origin of species*) where they express their opinions about the evolution of species.
- A description of deductive and inductive methods to obtain scientific knowledge which is taken from the book *About Science* by Barry Barnes.

Fig. 1: This is blog's initial page. Here we can found author's presentation and a piece of news based on science. The blog is written in Catalan language.

Students have different options to solve this exercise: to give their personal opinion about some of the texts' aspects, to contextualize the society in which it was produced, to search information about its author, etc. The professor evaluates the precision and rigor of the answer, its originality, the correction of the language and the way students express their ideas, and the number and quality of the references.

There are other pages in the blog apart from those which contain the texts. These pages try to connect science, their personalities and its activity with the rest of society. The contents of these pages are periodically renewed in order to give students a great variety of topics to consult and comment. These commentaries are voluntary but they have a positive influence in students' qualifications. Let's see which other sections we can find in the blog:

- *Art and science.* Here we can find a pictorial or sculptural representation with some relation with science and its history (See fig. 2). For example, until now there have been exposed pictures from classic authors such as *The ambassadors* by Holbein or *The anatomy lesson of Dr. Tulp* by Rembrandt. Illustrations and frontispieces of some scientific books have also been shown, as the one of the well known *De humani corporis fabrica* by Vesalius. As we can see, all of them are representations with many scientific connotations which can broaden students' minds to the social transcendence of scientific activity and its real repercussion in the context of the society in which this activity was produced. Students have to find the title of the work and its author. Normally, they find both of them very quickly. Then, they can comment it and analyse all the aspects related to science they are able to discover trying to contextualize them in its time.

- *The character.* In this page we find a picture of one or various personalities of the history of science and technology. The same way as before, pupils can comment everything they like about this person and his/her contributions to science or technology. In this academic year, 2010-2011, there has been represented pictures or photos of Carl von Linnaeus, Florence Nightingale, Eudald Carbonell[5], or the physicists who were present at the first Solvay Conference in 1911.

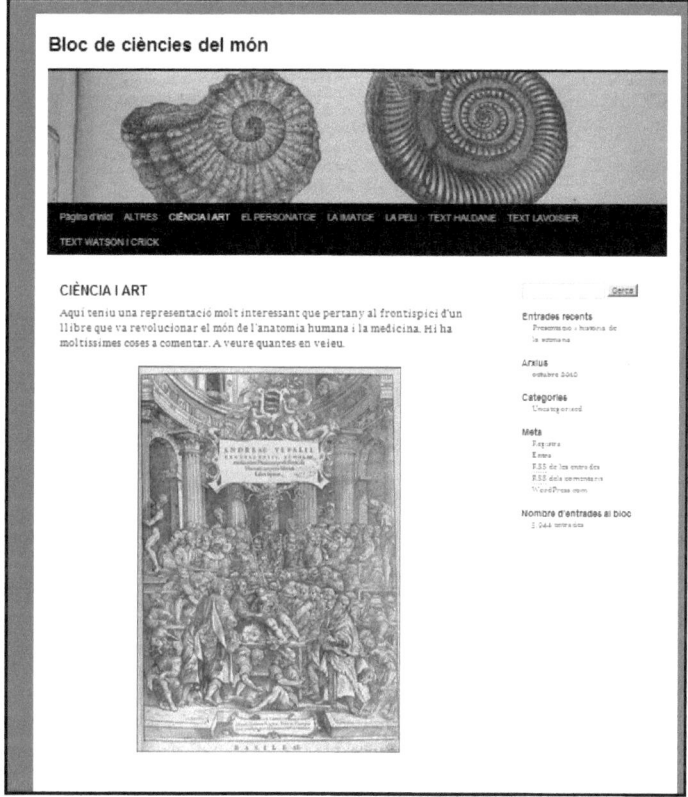

Fig. 2: An example of *Art and science* page, with the frontispiece of Vesalius' book.

- *The image*. In this section we will find whatever image with some relation with a scientific activity commented at class (See fig. 3). We can find from the frontispiece of *The sacred theory of the earth* by Thomas Burnett to the image of a X-ray diffractometer, and also Lucy's skeleton, and so on. Once again, the objective is to provoke a reaction in students, make them search about science and discover scientific topics they couldn't have imagined in other ways.

Fig. 3: Lucy's skeleton as it was shown in *The image*.

- *The film.* Another way to appreciate how non specialist people are interested in science is to verify how scientific activity has its place in commercial movies. There is a significant number of them whose plot is the history of science or its projection in the future. In any case, watching this films make pupils reflect on how science has a direct repercussion in people's life and, frequently, the ethical problems which can arise. Topics like the origin of humankind, the technological future, or the science-religion debate are the main theme of many movies. Some examples of these type of movies would be

2001 A space odyssey by Stanley Kubrick, *La guerre du feu* by Jean-Jacques Annaud, *Blade Runner* by Ridley Scott or *Agora* by Alejandro Amenábar. This section directs the students to watch quality films which can introduce them to an interesting and formative message.

The blog has another page entitled *Others* where students have the possibility to introduce any text, photo, film, etc. that they consider interesting to share with their classmates. The initial page of the blog introduces every week a scientific innovation that the teacher considers suitable to generate a debate.

At the present time, the evaluation about students' use of the blog is very positive. Their criticisms have been good enough and the number of entrances has exceeded the best expectations of the teacher. Since the blog was published three months ago, more than 5.000 entrances have been received and students, in a number of sixty, have made more than 500 commentaries that have been published by the teacher after being read and evaluated. Moreover, the rhythm of entrances has been increasing as the course advances. Students' comments have improved too. This demonstrates the positive function of this kind of activity. At first, there was very common to find contributions "copy and paste", so frequent in the work of teenagers and, unfortunately, in University students or postgraduates who are supposed to be better formed. The teacher's task must try to change this habit and promote students' personal opinion and, in any case, teach a rational use of the research, always explaining the resource which has been consulted.

The coming of a new methodology in the teaching of a subject must be supported by a non traditional system of evaluation. Consequently, written exams that only evaluate knowledge learned by heart coming from a textbook or the topics explained by the teacher are substituted by other methods which take into account other abilities. For example, it is very important to evaluate how students are able to search information, write a commentary of any topic, work in a team, or present their results to their classmates. Their aptitudes in using ICT are also of vital importance in our knowledge society. All these considerations must be reflected in the system of evaluation. Even more in a subject like SCW which has the objectives we have explained before.

In the case we are describing here, 40% of final qualification is obtained with students' participation in online activities, writing commentaries to their professor's proposals in the blog. This important part is given by the number and quality of these comments. This way, the consulting of the blog, the ulterior search for information, the elaboration of commentaries with their pertinent references to other resources found on the internet and the opinion of students have a significant importance in their final qualification. The remaining 60% is divided into two different blocks. In the first one, 30%, the teacher values the capacity of students to express correctly in written and oral form. Students achieve this part of the qualification through brief exams, doing little tasks proposed by the teacher, and explaining parts of the curriculum to their classmates using digital presentations. The second block, the other 30%, is obtained by a multiple choice exam. This is a way to improve students' abilities in this type of exams, so frequent in the university. A minimum of qualification is required in every block to pass the subject. This is to avoid the abandon of one of them by the students, and to give the same importance to all parts of this evaluation method.

To end with, it is obvious that this experience must be in a continuous revision. In our case, this new subject is being explained using ICT, as the knowledge society demands, and incorporating the history of science and technology to its curriculum in a significant way. From now on, new methodologies should be used to make this subject more attractive and useful for students: MOODLE, creation of a web, use of social networks, etc. World changes and teaching methodology must change too.

Second experience: Using a webpage as a reference of classical texts of the history of science.

In 2007, the other author of this paper, Agustí Camós, designed a webpage to promote among his students the reading of short texts by scientists of other periods of history[6]. This webpage was built in order to facilitate the access of teachers and students to significant texts of the history of science. Then the professors could work them with their students of secondary education (Camós:2008). This web is intended for helping teachers who teach natural sciences in Catalonia to locate historical texts on science translated into Catalan. This limitation in finding these types of texts is consequence of the enormous difficulties that the Catalan language has overcome the last three centuries. In this period, this language has been official in the Catalan country in only a few decennia. As a cause of this, a great number of the classical texts of history of science which have been translated to many of the languages used in Europe either have not been translated into Catalan or they are very difficult to be found.

Not only texts by internationally known authors are included in this web, but also texts by Spanish and Catalan authors. For instance, important scientific topics as the theory of evolution are found in classical texts by Lamarck or Darwin, but also in other texts by local scientists who participated in the evolutionist debate in Catalonia. That's the case of Odón de Buen or Josep Fuset Tubià, both professors in the University of Barcelona in the beginnings of the 20th century. This is the way to facilitate the approximation to the particularities of the Catalonian science. Moreover, it makes the recovery of the texts of scientists who worked in Catalonia possible.

Translations of texts into Catalan language are accompanied by the original version. This fact let students work either in the translated version or in the original one. They can also verify how translations from one language to another may produce important difficulties and, sometimes, notable distortions. Other teachers of foreign languages, basically French, English and

Classical languages may use these scientific texts in their classes. Unfortunately, this is not very common.

The webpage is divided in some sections. At first, there is a brief presentation where its contents are explained. Then we can find an essential tool, a table where the next information appears: author, subject, topic, original language, original document and degree of difficulty (See fig. 4). The function of this table is to give the reader enough information to decide quickly whether a text is interesting for him or not. There we can also find links to go to any of the thirty-five texts.

Autor	Matèria	Tema	Idioma original	Document original	Dificultat
Aristòtil 1	Biologia	La generació espontània	grec	Història dels animals	Senzill
Aristòtil 2	Biologia	L'herència en el món clàssic	grec	Història dels animals	Senzill
de Buen	Biologia	L'evolució zoològica	Castellà	Història Natural	Senzill
Buffon 1	Geologia	Temps en la història de la Terra	Francès	Les Époques de la nature	Mitjà
Buffon 2	Biologia	Els canvis en les espècies	Francès	Histoire Naturelle	Difícil
Buffon 3	Biologia	El concepte d'espècie	Francès	Histoire Naturelle	Senzill
Buffon 4	Biologia	El concepte d'espècie	Francès	Histoire Naturelle	Difícil
Crick Watson	Biologia	L'estructura de l'ADN	Anglès	"Una estructura per l'àcid desoxiribonucleic"	Senzill
Darwin, Charles 1	Biologia	La lluita per la vida	Anglès	On the origin of species	Mitjà
Darwin, Charles 2	Biologia	Variabilitat i selecció natural	Anglès	On the origin of species	Mitjà
Darwin, Charles 3	Biologia	Domesticitat, variabilitat i selecció natural	Anglès	On the origin of species	Mitjà
Darwin, Charles 4	Biologia	Les Galàpagos	Anglès	Journal of researches	Senzill
Darwin, Erasmus	Biologia	Evolucionisme predarwinià	Anglès	The Temple of Nature	Mitjà
Demòcrit 1	Química	Teoria atòmica de Demòcrit	Llatí	De Finibus	Mitjà
Font i Quer 1	Biologia	Les plantes pluricel·lulars tal·lòfites més senzilles	Català	Iniciació a la Botànica	Senzill
Fuset 1	Biologia	Creacionisme i evolucionisme	Castellà	Manual de Zoologia	Mitjà
Haldane 1	Biologia	L'origen de la vida a la Terra	Anglès	"The Origin of Life"	Mitjà
Hooke 1	Biologia	La primera observació d'una cèl·lula?	Anglès	Micrographia	Mitjà
Lamarck 1	Biologia	Fixisme i evolucionisme	Francès	Philosophie zoologique	Senzill
Lamarck 2	Biologia	Teoria evolucionista de Lamarck	Francès	Philosophie zoologique	Mitjà
Lamarck 3	Biologia	L'evolució humana segons Lamarck	Francès	Philosophie zoologique	Mitjà
Lamarck 4	Biologia	El temps i l'evolució	Francès	Philosophie zoologique	Mitjà
Lavoisier 1	Química	Equació química	Francès	Traité élémentaire de chimie	Mitjà
Lavoisier 2	Química	L'objectiu de la química	Francès	Traité élémentaire de chimie	Mitjà
Linné 1	Biologia	Nova classificació dels vegetals	Llatí	Critica botanica	Mitjà
Lucreci 1	Química	Els quatre principis	Llatí	De Rerum Natura	Senzill
Lucreci 2	Química	La matèria feta per àtoms com les paraules per lletres	Llatí	De Rerum Natura	Mitjà
Marcet 1	Química	La química	Anglès	Conversations on Chemistry	Senzill
Margalef 1	Biologia	El naixement de l'ecologia	Català	Cent anys d'Ecologia	Senzill

Fig. 4: This is the table we can find in the web which indicates its contents with some information.

Under this table, the thirty-five texts can be found. Every text is organised in the same way:

1) Name of the scientist with his birth and death date.
2) Reference of the resource where the text comes from, with the name of the translator.
3) A portrait of the scientist.
4) The topic under discussion.
5) The Catalan version of the text.
6) The bibliographical reference of the original text.

7) The original version of the text.
8) Direct links to interesting webs which are related to the author or his work.
9) Other complementary data.
10) The link to the initial page.

We are not going to find excessive additional information in the web in order to make its use easier. That is the reason why there are not many links to other webs or a lot of complementary data. Only the most essential ones have been included.

When teachers start using this web, they will look at the initial table and there they will check which authors and topics are available and there degree of difficulty. Then they can choose the text which is better adapted to the necessities of their students to use it in a convenient way. The text can be used after having been printed or directly projected on a screen. It can also be used through the MOODLE platform, for example to promote a debate in a forum. The elaboration of a questionnaire depends on the context in which the text is going to be used.

In this moment the webpage contains twenty-five texts on biological topics, six texts on chemistry, two on geology, one on physics and one on maths. There are ten texts in French, nine in English, six in Latin and two in ancient Greek. Eight texts belong to the Greek and Roman periods, one to the 17th century, eight to the 18th century, twelve to the 19th century and six to the 20th century.

Even though this web only contains a limited number of texts, it has been used by a significant number of secondary professors in Catalonia. Our wish is to increase significantly the number of historical texts in all the topics, periods and cultures. Taking into account the increasing multiculturalism of our society, it would be interesting to add scientific texts coming from Non western cultures, such as Arabian, Chinese or Indian.

Some historians of science and teachers of secondary level have collaborated in the construction of the web. Only a collective task, including the work of historians and teachers and the collaboration of institutions, will guarantee the completion of this project.

Conclusions

The need to give basic knowledge about the nature of science and its transcendence in the present world has led Spanish Ministry of Education to design a new subject, *Science for a Contemporary world* (SCW) which must be taught to the students in the 1st year of baccalaureate (16-17 years old). Just from the beginning, the experts who designed its curriculum tried to avoid turning SWC into a simple enumeration and description of relevant scientific topics of our time. On the contrary, this subject should contribute to teach how science works and why scientific methodology is believed to be the best way to achieve knowledge.

In the opinion of the authors of this paper, the systematic use of the history of science and technology may improve the educational task in this subject. Real experiences of scientific activity in the past and the change of paradigmatic ideas through the history can contribute to explain the nature of scientific activity at the present time.

The use of the new information and communication technologies (ICT) is absolutely necessary in this subject. One of the sections of its curriculum is dedicated to the knowledge society and the technological development in our time. That is why the authors have designed an educational methodology based on the consult of a webpage or the participation in a blog designed by the teacher. The internet has proved to be an exceptional informational resource which, in some cases, can conveniently replace the textbook as a pedagogic tool. That is the case when available textbooks do not come up to teachers' expectations.

The first experience which has been explained is about a blog specifically designed for SCW. This blog shows, in different pages, scientific texts and images, great personalities, films that deal with science, scientific news, etc. Students must actively participate in the blog. They have to send comments expressing their opinions about some of the contents that the teacher has included in the blog. Then, the teacher publishes the commentary after being read and evaluated.

The second experience which is described in this paper is about a webpage which can be used in any scientific subject. In this webpage, some significant fragments of classic scientific books in their original version and translated into Catalan can be found. Its main objective is to make it possible for educators to find classical texts in an easy way and to promote students' reflection on the origin of scientific knowledge and its interpretation from the original text.

The use of the history of science and the new technologies in SCW has an obvious repercussion in the learning methodology. But it also has to transcend to the methodology of evaluation of this subject. The capacity of interpretation, making relations and oral and written expression which students should learn with this system must be evaluated by the teacher and have an important weight in the final qualification.

The authors believe that the use of the history of science as the backbone of SCW gives teachers an excellent tool to explain the scientific activity and all its consequences. Students will be more interested in the subject if information and communication technologies are commonly used. Moreover, in this way the subject is better integrated in the communicational systems of our society. The new challenge of teachers is to improve the contents of the subject through the search of new resources on the internet, and to use the new ICT tools which are constantly appearing in our changing world.

References

Barnes, B. (1985). *About Science*, Basil Blackwell Ltd, Oxford.

Camós, A. (2008) "Un pas en la construcció d'una potent eina informàtica d'història de la ciència en català per a l'ensenyament secundari" *Ciències*, 10, p 7-11, http://www.raco.cat/index.php/Ciencies/article/view/105001/131294 (accessed 10 March 2011)

Ezquerra, A. (2010). "Ciencias para el mundo contemporáneo y comunicación audiovisual", *Alambique* n. 64, pp. 59-71, http://alambique.grao.com/creditos/ficha_articulo.asp?id=6273 (accessed 10 March 2011)

Izquierdo, M. et Al. (2006). "Relación entre la historia y la filosofía de las ciencias II", *Alambique* n.48, http://alambique.grao.com/creditos/ficha_articulo.asp?id=6015 (accessed 10 March 2011)

Mas, V. (2006). "Ciencias para la ciudadanía en Francia. Un análisis de la propuesta francesa similar a la que va a ser introducida en España.", *Alambique* n. 49, pp. 30-41, http://alambique.grao.com/creditos/ficha_articulo.asp?id=6024 (accessed 10 March 2011)

Millar, R.; Hunt, A. (2006). "La ciencia divulgativa. Una forma diferente de enseñar y aprender ciencia", *Alambique* n. 49, http://alambique.grao.com/creditos/ficha_articulo.asp?id=6023 (accessed 10 March 2011)

Pedrinaci, E. (2006). "Ciencias para el mundo contemporáneo: ¿Una materia para la participación ciudadana?", *Alambique* n. 49, http://alambique.grao.com/creditos/ficha_articulo.asp?id=6022 (accessed 10 March 2011)

Textbooks

Anguita, F. et Al. (2008). *Ciències per al món contemporani*, Grup Promotor Santillana, Barcelona.

Cuadros, J. et Al. (2008). *Ciències per al món contemporani*, Ed. Castellnou, Barcelona.

Fornells, M. et Al. (2008). *Ciències per al món contemporani*, Ed. Casals, Barcelona.

González, M. et Al. (2008). *Nexus. Ciències per al món contemporani*, Pearson Educación, Madrid.

Grau, R. et Al. (2008). Ciència en context. Ciències per al món contemporani, Ed. Teide, Barcelona.

López, A. et Al. (2008). *Ciències per al món contemporani*, Ed. Mc Graw Hill, Madrid.

Pedrinaci, E. et Al. (2008). *Ciències per al món contemporani*, Ed. Cruïlla, Barcelona.

Rubio, N. et Al. (2008). *Ciències per al món contemporani*, Ed. Barcanova, Barcelona.

---

[1] From many years ago, the use of history of science and technology as a vertebral axis of science learning has been an object of study. For instance, the Catalan Society of His-

tory of Science and Technology (SCHCT) has organised six conferences on Learning and History of Science. In these conferences, a high number of educators have showed their experiences. Many of them have been published in the *Actes* (Journal of the SCHCT).

[2] The references of these textbooks can be found in the bibliography.

[3] Ignaz Semmelweis (1818-1865) was a Hungarian physician who discovered that puerperal fever that affected women after the childbirth in Vienna's hospital could be drastically eradicated when doctors disinfected their hands before making the obligatory revisions.

[4] Its URL is http://rsucarrats.wordpress.com/ . (accessed 10 March 2011)

[5] Eudald Carbonell (1953-) is a very well known Catalan palaeontologist. He has created an analysis system of prehistoric technology that has been applied, for instance, in the Sima de los Huesos in Atapuerca (Burgos, Spain) where *Homo antecessor* has been discovered and described.

[6] It can be found in the address http://www.xtec.cat/~acamos/ (accessed 10 March 2011) entitled: the great scientists in Catalan and in his original language in the secondary education.

The Importance of Games of Chance at the Inception of Probability Theory

Fàtima Romero Vallhonesta

Departament de Matemàtica Aplicada I, Universitat Politècnica de Catalunya, Barcelona;
fatima.romerovallhonesta@gmail.com

ABSTRACT: The history of mathematics provides enough elements to justify its inclusion in the secondary school syllabus, as evidenced by the historical contexts introduced in the new syllabus published in Catalonia in June 2007. The current legislation includes a non-exhaustive list that contains, among others, the title: "The Birth of Probability Theory". In this paper we present some ideas for developing this context on the basis of two historical problems, the first dealing with the most probable sum when throwing two or three dice, and the second with the problem of dividing the stakes equitably when a fair game is halted before the end.

Introduction

In 2007 a new syllabus[1] was published in Catalonia giving guidelines for teaching activity in primary and secondary schools. A non-exhaustive list can for the first time be found in the secondary school syllabus, with the possibility of enlarging it, through historical approaches related to the contents at every academic level. For the third year (thirteen-year-old pupils), for example, the proposals are as follows:

The origins of symbolic algebra (Arabic World, the Renaissance)

The relation between geometry and algebra and introduction to Cartesian coordinates

The geometric resolution of equations (Greece, India, Arabic World)

The use of geometry to measure the distance Earth -Sun and Earth- Moon (Greece)

The birth of the theory of probability

As a member of the History of Mathematics Group[2] of the Barcelona Association for the Study and Learning of Mathematics (ABEAM), I have contributed to a development of historical materials for use in the classroom[3]. These materials consist of activities related to some historical topics and linked to the curriculum.

The aims of this group are: firstly, to provide teachers with a brief history related to different subjects in the curriculum, secondly, to give extensive bibliography about the subjects involved and its teaching methods; thirdly, to provide the education community with translations of outstanding historical texts and finally, to draw up activities for students. In these activities, students are required to search for information, read texts, and answer questions aimed at helping them understand the demonstration or the reasoning involved in the historical text.

A further aspect highlighted in the new syllabus is the acquisition of key competences, the learning of which is recommended in an interdisciplinary way. Emphasis is also given to the importance of the logical thinking, which while not exclusive to mathematics is certainly one of its characteristics.

The activities presented here are designed according to the guidelines of the Catalonian syllabus, and thus take into account the importance of acquiring key competences. These activities for the classroom are related to two historical problems[4], the first dealing with the most probable sum when throwing two or three dice, and the second with the problem of dividing the stakes equitably when a fair game is halted before the end. For the first problem, we suggest an activity developed from the reading of a fragment of an opuscule from Galileo Galilei (1564-1642). For the second problem, the activity we propose involves the formulations of the problem by Luca Pacioli (1445-1517), Niccolò Fontana ("Tartaglia") (1500-1557) and Girolamo Cardano (1501-1576), and its solution contained in an anonymous 15th century manuscript, as well as the solution given by Blaise Pascal (1623-1662) in his correspondence with Pierre de Fermat (1601-1665).

First Activity. Historical Text: *Concerning a Discovery about Dice*

The first activity we propose involves the last context included in the curriculum for third-year students, which concerns the most probable sum when throwing three dice, and is based on a brief treatise by Galileo. Galileo Galilei (1564-1642), was an Italian physicist, mathematician, astronomer, and philosopher who played a major role in the Scientific Revolution. His achievements include improvements to the telescope and consequent astronomical observations and support for Copernicanism. Galileo's work in physics helped greatly to make experimental measurements and mathematical calculations more prominent in all the sciences. Galileo was one of the founders of modern science. Frequently Galileo is singled out as the most pivotal of these founders and called the Father of Modern Science. His contribution to a development of probability theory is, perhaps, his less known work.

Galileo's text, which is undated and untitled, can be found in the 8th volume of the national edition of Galileo's works. In the last part of this volume, Antonio Favaro, editor of Galileo's works, includes texts the date of which has proved impossible to determine. One of these Galileo's texts, entitled *Sopra le scoperte de i dadi*[5] by Favaro, deals with the problem of finding the most probable addition when throwing three dice, and was published for the first time in Florence in 1718 together with other works on probability.

It seems that Galileo wrote this opuscule in answer to a question that someone posed to him, and which refers to the most likely situation for obtaining a sum of 10 rather than a sum of 9 when throwing three dice, although the number 9 and the number 10 can be obtained in six different ways.

Galileo begins his booklet by explaining that in a game of dice the fact that certain numbers are more advantageous than others depends on the ways in which each number can be made up. He gives some examples of numbers that can only be made up in one way, 3 and 18: the first can only be obtained with three "ones" and the second only with three "sixes". He also gives examples of numbers that could be obtained in several ways; that is, a 6 with the combinations (1,2,3), (2,2,2) and (1,1,4), and a 7 that can be obtained with the combinations (1,1,5), (1,2,4), (1,3,3) and (2,2,3). In reply to a question that someone asked him about the fact that although 9 can be made up in as many ways as 10, long observation had led dice-players to consider 10 more advantageous than 9, Galileo said that he would expound his point of view in the hope not only of solving this question but of opening up the way to a better and more detailed understanding of the reasons behind the game.

His analysis of the game starts with the throw of one die, which can lead to only six different results, since a die has six faces. Galileo goes on to consider what would happen if a second die were to be thrown. In this case, he says that 36 different results could obtained, since each face of the first die can be combined with each face of the second, and the number of such combinations is 6 times 6. Furthermore, if a third die is thrown, 216 different results could be obtained, since each face of the third die can be combined with each one of the 36 combinations of the other two dice, the number of such combinations being 6 times 36. If we take into account that the results that could be obtained by throwing three dice are 16, that is, from 3 to 18, the 216 combinations would be divided among these numbers. Thus it follows that many throws must correspond to some of these numbers, so if we know how many throws correspond to each number, we will find out what we wish to know.

After analyzing with concrete examples how many ways may lead to some results, Galileo states what he considers to be the three fundamental points; that is, three equal numbers can only be obtained in one way; two equal numbers and the third different number in three ways, and three different numbers in six ways. He then draws up the following chart (Fig. 1):

10		9		8		7		6		5		4		3	
6.3.1.	6	6.2.1.	6	6.1.1.	3	5.1.1.	3	4.1.1.	3	3.1.1.	3	2.1.1.	3	1.1.1.	1
6.2.2.	3	5.3.1.	6	5.2.1.	6	4.2.1.	6	3.2.1.	6	2.2.1.	3				
5.4.1.	6	5.2.2.	3	4.3.1.	6	3.3.1.	3	2.2.2.	1						
5.3.2.	6	4.4.1.	3	4.2.2.	3	3.2.2.	3								
4.4.2.	3	4.3.2.	6	3.2.2.	3										
4.3.3.	3	3.3.3.	1												
	27		25		21		15		10		6		3		1

Fig. 1: Reproduction of Galileo's table as can be seen in the page 5 of his booklet

With the points from 10 to 3 along the top row, the different triples making up these points below, and next to these triples the number of ways that each triple can be obtained. The last row gives the sums of these ways.

According to this analysis, Galileo says that it is possible to understand why a sum of 10 can be obtained more frequently than a sum of 9, since the first sum can be made up in 27 ways and the second in 25 ways. He goes on to add that it is enough to focus the analysis on numbers between 3 and 10, because the higher sums of points, from 11 to 18, can be obtained by a number of combinations similar to the eight sums shown in the table.

We believe that this is a suitable text for an introduction to probabilistic reasoning. Galileo's explanations are very clear, and his intention not only of answering the question that was raised to him, but also of understanding the logic of the game, can help students to understand the scientific method involved.

It is important to remark that we have yet to address the issue of probability[6], and have so far only dealt with frequency, which is a much easier concept.

The First Activity for the Students

This activity is intended to contribute to the acquisition of key competences by students, particularly in achieving mathematical competence, the learning-to-learn competence, the competence of communication in the mother tongue and the interpersonal and civic competence, and can be also useful to introduce the probability concept in the classroom. The activity should be conducted in pair-work groups consisting of three students throughout a one-hour session. Students are required to look for information about Galileo beforehand at home.

- ✓ Look for information about Galileo-Galilei: his writings, his contributions to physics and astronomy, his condemnation as a suspected heretic by the Catholic Church and about the historical circumstances of Galileo's time.
- ✓ If you throw one die, how many different results can be made up?

How many if you throw two dice?

- ✓ How many if you throw three dice? Do all the results that can be obtained have the same chance of being made up? Justify your answer.
- ✓ Read Galileo's text about the dice game.
- ✓ In this text, how does Galileo justify that there are 216 possibilities when throwing three dice?
- ✓ Justify in Galileo's way that the sum of 10 is more frequent than 9 in the game that consists in throwing three dice and adding up the results, which is actually a fact that the most experienced players already knew.
- ✓ Complete Galileo's chart with the results from 11 to 18.

Second Activity. Historical Texts about the Division of the Stakes in an Interrupted Game.

Apart from the activities for pupils, some additional readings are also recommended for teachers. In the case of the problem of the division of the stakes, an interesting introduction to the first calculus on probabilities can be found in Hacking, 1995 and in De Mora, 1989, and about the history of this problem prior to Pascal in Coumet, 1965. The problem concerns a game of chance involving two players who have an equal chance of winning each round. The stakes are the same for both players, who agree in advance that the first player to win a certain number of rounds will collect the entire amount. We then have to assume that the game is interrupted by external circumstances before either player has achieved victory. The question is how the stakes can be divided fairly.

This is a problem that had been addressed by several mathematicians who made different contributions that eventually led to a satisfactory solution, which was finally provided by Pascal in his correspondence with Fermat. It is an interesting problem to study in the classroom, because the construction of knowledge is evident and the importance of group-work can readily be appreciated. Although there was no real group working on this problem, the different opinions and solutions given by different authors were used as a starting point for other authors who improved these solutions, all of whom contributed to the correct solution.

Many authors dealt with this problem of dividing the stakes, also known as "the problem of points", which can be found in Pacioli's *Summa de Arithmetica, Geometria, Proportioni et Proportionalita* (Fig. 2), although it emerged prior to this author, since he says that he had found different opinions in one sense or another, all of which appeared incoherent to him. In the solution given by Pacioli, each player receives a share of the stake proportional to his number of points as against the total number of points attained when the game was interrupted. This solution was later criticized by authors such as Cardano and Tartaglia, who gave alternative solutions that despite containing some mistakes represented a step forward to the right solution, which finally arrived in 1654 with the contribution by Pascal, who also dealt with the "problem of points" in his work *Traité du triangle arithmétique*.

This thread is very interesting didactically speaking, but prior to Pascal there was a lesser known solution in a manuscript dating from the first half of the 16th century (Franci, 2002), whose author was probably an abacus master. This provides a very interesting text for analysis by students.

Luca Pacioli's Solution

In the *Summa,* the problem is formulated as follows:

Two teams play a ballgame in such a way that a total of 60 points is required to win the game, and each goal is worth 10 points. The stakes are 10 ducats. An incident occurs that prevents the game from reaching its finish, at which time one side has 50 points and the other 20. One wants to know what share of the stake belongs to either side (Pacioli, 1494, 197r)[7].

Fig. 2: Index of the *Summa*

Pacioli states that he has already found different opinions on this question (thus the problem had been addressed prior to Pacioli), all of which appear incoherent to him in their argument, and goes on to say that he will state the truth of the matter.

Pacioli solves the problem in three ways, all of them based on the points accumulated by the teams, regardless of the result they might have obtained if the game had reached its conclusion. All these three ways give the same result for the players' share of the stakes, and below we give the easiest of these ways:

Pacioli asked how many goals at most are the two teams together able to score, which is 11; that is, when each team has 50 points, so who wins next will be the winner of the game. The proportion of these 11 goals corresponding to the first team is $\frac{5}{11}$ and the proportion for the second team is $\frac{2}{11}$.

The sum of this is $\frac{7}{11}$, so $\frac{7}{11}$ is worth 10 points. Pacioli then asked what would be the outcome with $\frac{5}{11}$ and $\frac{2}{11}$, and the answer is $7\frac{1}{7}$ of 10 to the team with 50 points and $2\frac{6}{7}$ of 10 to the team with 20.

In the end, each team will receive a share of the stakes proportional to its number of points as against the total number of points achieved, so the maximum number of goals that may have been scored by both teams, and which Pacioli took into account to find the solution, are irrelevant.

Cardano' Contribution

The problem of a division of the stakes can also be found in the *Practica arithmetice et mensurandi singularis* by Girolamo Cardano (Fig. 3). This author criticizes Pacioli's solution because he fails to take into account the number of games remaining to each player in order to win the game. However, the solution that Cardano suggests, although it works for Pacioli's example, is not valid in general.

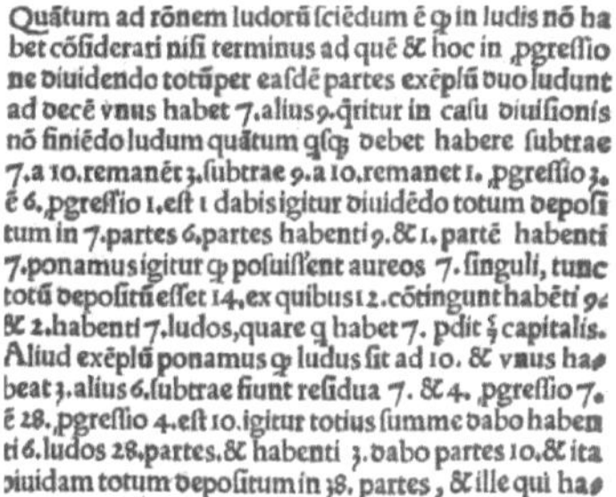

> Quātum ad rōnem ludorū fciēdum ē q̄ in ludis nō habet cōfiderati nifi terminus ad quē & hoc in ‚pgreffione diuidendo totū per eafdē partes exēplū duo ludunt ad dece vnus habet 7.alius 9.q̄ritur in cafu diuifionis nō finiēdo ludum quātum q̄fq̄ debet habere fubtrae 7.a 10.remanēt 3.fubtrae 9.a 10.remanet 1. ‚pgreffio 3. ē 6. ‚pgreffio 1.eft 1 dabisigitur diuidēdo totum depofitum in 7.partes 6.partes habenti 9. & 1. partē habenti 7.ponamus igitur q̄ pofuiffent aureos 7. finguli, tunc totū depofitū effet 14,ex quibus 12.cōtingunt habēti 9. & 2.habenti 7.ludos,quare q̄ habet 7. pdit ⅔ capitalis. Aliud exēplū ponamus q̄ ludus fit ad 10. & vnus habeat 3.alius 6.fubtrae fiunt refidua 7. & 4. ‚pgreffio 7. ē 28. ‚pgreffio 4.eft 10. igitur totius fumme dabo habenti 6.ludos 28.partes,& habenti 3. dabo partes 10. & ita diuidam totum depofitum in 38. partes , & ille qui habet 3.perdit ⁸⁄₁₃ fui capitalis.

Fig. 3: The formulation of the problem in *Practica arithmetice*

Nevertheless, Cardano's contribution is very important because it points to a new direction for finding the solution to the problem.

The situation that this author proposes is a game in which two players gamble up to ten points and the stakes are 7 golden pieces each. One player has 7 points and the other 9. The question is how much each player would have if the game were to be interrupted and the stakes divided. Cardano's method[8] consists in subtracting 7 and 9 from 10 and considering the remainders 3 and 1. To set a proportion, the author considers the sum of the progressions[9] of 3 and 1, which is 7, and then by dividing the total into 7 parts, 6 of these parts, which is 12 golden pieces, will be awarded to the player having 9 points, and 1 part, which is 2 golden pieces, to the one having 7 points. Cardano's reasoning concerns the difficulty of winning should the game continue. He takes into consideration not only the points that each player has won, but also the points that they would need to finish the game. This is an important point for further reasoning.

Tartaglia's Contribution

Fig. 4: The Tartaglia's 2nd part of *General Tratatto*

Tartaglia deals with this problem in his *General Trattato di Numeri et Misure* (Fig. 4), explaining Pacioli's solution and rejecting. He bases his rejection on the fact that should the game be interrupted when one of the players has won 10 points, for example, and the other one 0, then according to Pacioli's method, proportional to the points obtained, the first player would obtain the whole amount while the other player would receive nothing, which is not fair to him. Before his proposal, Tartaglia expresses his scepticism about finding the correct right solution to such a problem, which he considers a legal issue rather than a question of mathematical reasoning. What he proposes is that the player who has the advantage gets his wager plus a part of the remainder that is proportional to his advantage. For example, in the problem arising from Pacioli, if one team has 10 points and the other 0, the team that has scored 10 will have a sixth of the complete game and should receive their wager plus a sixth part of the money that the other team has staked; that is, if the stakes are 22 ducats per team, the one with 10 points should receive 22 ducats plus $\frac{22}{6}$ of 22; that is, $25\frac{2}{3}$ ducats, while the other team should receive $18\frac{1}{3}$ ducats.

Pascal's Solution. Pascal's Correspondence with Fermat

The problem of division was proposed to Pascal and Fermat, probably in 1654, by the Chevalier de Méré, a gambler who is said to have had an unusual ability "even for mathematics", and Pascal and Fermat went to address this problem in their correspondence.

This correspondence between Fermat and Pascal was vital for the development of modern concepts of probability, and it is unfortunate that Pascal's introductory letter to Fermat is no longer extant.

In letter LXX to Fermat, Pascal tackles the problem of the division of the stakes in an unfinished game, and explains how he set about determining the value of each of the shares when two gamblers play a game of dice in three throws, for example, in which each player has wagered 32 pistoles:

> Let us suppose that the first of them has two (points) and the other one. They now play one game of which the chances are such that if the first wins, he will win the entire wager that is at stake, namely, 64 pistoles. If the other wins, they will be two to two, and in consequence, if they wish to separate, it follows that each will take back his wager, namely 32 pistoles each.
> Consider then, Monsieur, that if the first wins, 64 will belong to him. If he loses, 32 will belong to him. Then if they do not wish

to play this point, and separate without doing it, the first should say "I am sure of 32 pistoles, because even a loss gives them to me", but as for the other 32, perhaps I will have them and perhaps you will have them, the risk is equal. Therefore let us divide the 32 pistoles in half, and give me the 32 of which besides I am certain." He will then have 48 pistoles and the other will have 16. Now let us suppose that the first has two points and the other none, and that they are beginning to play for a point. The chances are such that if the first wins, he will win the entire wager, 64 pistoles. If the other wins, behold, they have come back to the preceding case in which the first has two points and the other one.

But we have already shown that in this case 48 pistoles will belong to the one who has two points. Therefore if they do not wish to play this point, he should say, "If I win, I shall gain all, which is 64. If I lose, 48 will legitimately belong to me. Therefore, give me the 48 that are certain to be mine, even if I lose, and let us divide the other 16 in half because there is as much chance that you will gain them as I will." Thus he will have 48 and 8, which are 56 pistoles.

Suppose finally that the first has only one game and the other none. You will see, Monsieur, that, if they begin a new game, the condition is such that, if the first wins it, he will have two games to nothing, and hence, by the preceding case, 56 belongs to him; if he loses, they are one game to one game: so 32 pistoles belong to him. Therefore he must say: "If you do not wish to play any longer, give me 32 pistoles which are certainly mine, and let us divide the other 56 in half. From the 56 remove 32; 24 remain; divide therefore 24 in half, you take 12 of them and I take 12, which, together with 32, make 44[10]

The solution given by Pascal takes into account the games remaining to win. He does not calculate from the possibilities of winning, but from the games that will definitely be won. The method that he uses consists in starting from the end and going back step by step to the first game (Fig. 5).

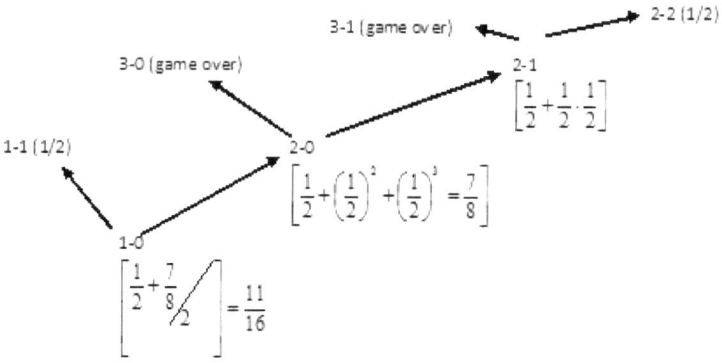

Fig. 5: Our scheme to help the understanding of the Pascal's solution

The Solution in the Urbinati Latini manuscript 291

The manuscript *Ur. Lat.* 291 from the Biblioteca Apostolica Vaticana is anonymous and undated, but according to Van Egmond it can be situated in the first half of the 15th century. In this case the wording of the problem refers to 3 players, who we will call A, B and C and who play up to 3 games. The stakes are 8 *denari* each and the game should be interrupted when A has won 2 games, B, 1 game and C 0. As Franci (Franci, 2002, 256) suggests, we express this situation as <2,1,0> to make the explanation easier. The resolution starts by considering what would happen if A and B had 2 games and C only 1. In this case, according to the author, if C won the following game, a third of the total amount would belong to each player. So if C wins, he will receive $\frac{1}{3}$ of the wager and if he loses he receives nothing. Therefore C has to take $\frac{1}{3}\left(\frac{1}{3}+0+0\right)$, that is $\frac{1}{9}$ of the stakes, while A and B have to take the half of the remainder, that is, $\frac{4}{9}$ each. After that, the author studies the cases <2,1,1>, <2,2,0>, <2,1,0> and <2,0,0> with very clear reasoning and gives the correct solution for each of them.

The Second Activity for the Students

Probability should be introduced into the classroom prior to this activity, and students should do some exercises to practice and familiarize themselves with the probability concept. To do this activity students have to work in group.

The activity starts by posing the following questions to the students:
- ✓ Look for information about Luca Pacioli and the mathematics of his time, focusing on his main work *Summa de Arithmetica, Geometria, Proportioni et Proportionalita*.
- ✓ Read Luca Pacioli's problem about the division of the stake if the game is interrupted (we give the students the original text and our Catalan translation) and solve it.

The core of this activity is the problem we analyzed above about the division of the stakes in an interrupted game. Students are given the wording of Pacioli's problem without knowing that this is a probability problem and are required to solve it individually. They then have to discuss their solutions together in groups of three or four. The aim is for students to become familiar with the problem before they are given different the different solutions that have been proposed throughout history.

After that, they are given Pacioli's solution with the following questions:
- ✓ Do you think it is logical to distribute proportionally the points that each player has at the moment that the game is interrupted?
- ✓ Should the game finish; is it possible for the player with the least number of points at that moment to be the winner?
- ✓ What were the conditions of the game?
- ✓ Would it be better to take into account what would have happened if the game had finished?
- ✓ Write all the possibilities for the game to finish.

On completion of these questions, the students are presented with the objections by Cardano and Tartaglia to Pacioli's solution, Pascal's solution and the first step towards the solution of the case given above from the *Urb. Lat.* 291 manuscript.

- ✓ Look for information about Tartaglia's and Cardano's works.
- ✓ Read Cardano's and Tartaglia's objections to Pacioli's solution and comment on them.
- ✓ Follow the steps of Pascal's reasoning in the case where each player has bet 50 euros:
    1. If a player A has 2 points and player B has 1, what amount belongs to player A if the game is interrupted?
    2. What if player A has 2 points and player B still has no points?
    3. What if A and B had 1 point each?
    4. Finally, what would happen if A had 1 point and B had no points when the game was interrupted?
- ✓ In the first step to the solution in the manuscript, the amount that each player should take is obtained by reasoning from the point of view of the player with only one game. Repeat the reasoning from the point of view of one of the players with two games and confirm that you obtain the same result.

Final Remarks

As we have concluded from the results of the activities we have undertaken, introducing concepts and procedures from historical activities into the classroom has been a highly enriching experience for both students and the teachers who proposed them. In these kinds of activities, students are required to develop skills that are not called for in more standard activities, as well as putting their knowledge into practice, often without any context and in complex situations.

Such types of activities are aimed at providing students with greater learning of key competences. Unlike standard problems which, once classified, have an immediate solution, these activities stimulate the reasoning power of students. It is also important for students to acquire an appreciation of teamwork. In the case of the second activity, the contributions of different authors, although not as part of a team, helped to arrive at a satisfactory solution.

These activities have yet to be tried out in secondary schools[11], although the problem of the division of the stakes was proposed to the in-service teachers who attended an online course on History of Science[12]. As a tutor of 10 of these teachers, it was my task to comment on the solutions they proposed to this problem. Two of these teachers proposed a division like the one Pacioli advocated, while one gave two different solutions: the first in the case where we take into account the points that each player has won at the moment the game was interrupted (which corresponds to Pacioli solution), and the second solution in the case where we take into account the advantage that one of the players has over the other (which corresponds more or less to the Tartaglia's solution). Finally, another teacher also proposed two different solutions; the first was similar to Pacioli's, while for the second he said that this would be the solution in the case where the number of games remaining to finish would be taken into account. It is necessary to remark that the last two teachers made no comments about which of the solutions they suggested was the correct one.

Finally, we have to point out that not all historical texts are appropriate to work with secondary school students. A good balance between the historical texts and mathematical concepts in the curriculum is necessary, in a way that historical texts contribute to the understanding of the mathematical concepts and also to the improvement of the students' reasoning.

References

Coumet, E. (1965), « Le problème des partis avant Pascal», *Archives internationales d'historie des sciences*, 18:73, 245-272.

Cardano, G. (1539) *Practica Arithmetice et Mensurandi singularis*. Bernardini Calusci, Milan. Reprinted in Opera Omnia, Vol. 4, 1663. Chapter LXI, 143r-144r

De Mora Charles, M. (1989), *Los inicios de la teoría de la probabilidad. Siglos XVI y XVII*. Editorial Ellacuría, S.A.L. Erandio.

Demattè, A. (2006), *Fare matematica con i documenti storici. Volume per l'alunno*, Editore Provincia Autonoma di trento. IPRASE del Trentino, 133-137.

Egmond, W. van (1980), *Practical mathematics in the Italian Renaissance: a catalog of Italian abacus manuscripts and printed books to 1600*, Firenze, Istituto e Museo di Storia della Scienza.

Fermat, P. (1894), *Œuvres de Fermat*, Volume 2, 288-314. Letter LXX from Pascal to Fermat, Wednesday, 29 July 1654.

Franci, R. (2002), "Una Soluzione Esatta del Problema delle Parti in un Manoscritto della Prima Metà del Quattrocento", *Bollettino di Storia delle Scienze Matematiche*, vol. XXII, fasc 2, 253-265.

Galilei G. (1718), "Sopra le scoperte de i dadi", in *Le Opere di Galileo Galilei*, Edizione Nazionale (E.N.) di A. Favaro, Firenze, Vol. VIII, 591-594.

Hacking, I. (1995), *El surgimiento de la probabilidad*, Ed. Gedisa, Barcelona, 68-77.

Massa, M.R.; Romero, F.; Guevara, I. (2007), Teaching Mathematics through History: some trigonometric concepts, in *Proceedings of the 2nd International Conference of the European Society for the History of Science (ICESHS)*, Cracow, 150-157.

Pacioli, L. (1494), *Summa de Arithmetica, Geometria, Proportioni et Proportionalita, Distinctio nona*, tractatus I, 197, Paganino de Paganini, Venecia.

Pascal, B. (1954), *Oeuvres complètes*, Gallimard, Paris.

Romero,F.; Puig-Pla, C.; Massa, M.R.; Guevara, I. (2009), La trigonometría en els inicis de la matemática xinesa. Algunes idees per treballar a l'aula, in *Actes d'Història de la Ciència i de la Tècnica*. Nova época. Volum2 (1), p. 427-436.

Tartaglia, N. (1556), *General Tratatto di Numeri et Misure*, book 16, section 206. Venezia.

---

[1] DOGC (Official Daily Bulletin of the Generalitat of Catalonia), 4915, June 29th 2007

[2] The coordinator of the group is Mª Rosa Massa Esteve, and the other members of group are: Mª Àngels Casals Puit (IES Joan Corominas), Iolanda Guevara Casanova (IES Badalona VII), Paco Moreno Rigall (IES XXV Olimpíada), and Carles Puig Pla (UPC). The group is subsidized by the University of Barcelona *Institute of Science of Education* (ICE).

[3] As examples of these materials, see (Massa et al., 2007) and (Romero et al,. 2009).

[4] More activities about these historical problems can be found in (Demattè, 2006).

[5] We use the 1989 Catalan translation by the Philosophy Group of the "Casal del Mestre" in Santa Coloma de Gramenet. The Philosophy and History of Science Group of Santa Coloma is composed of secondary school teadhers in Catalonia. Since its establishment in 1986, this group has been a working group at the Casal del Mestre de Santa

Coloma and in has in his balance hundreds of publications, conferences, and issues of the journal of opinion.

[6] Later in the course, when we deal with probabilities, we can refer to Galileo's text again. Students could then calculate the probability of obtaining a sum of 9 and a sum of 10 when throwing three dice. They will also have to say whether they think this difference could be noticed only by observation, as Galileo suggested.

[7] Una brigata gioca apalla a. 60. el gioco e. 10. p caccia e fano posar duc. 10. acade p certi accideti che no possano fornire e luna pte a. 50. e laltra a. 20 se dimanda che tocca p pte. de la posta.

[8] A deep analysis of the Cardano solution can be found in (Coumet, 1965)

[9] In this text, by the expression *progression of n*, Cardano means the sum of the natural numbers up to n, that is, $\frac{n(n+1)}{2}$

[10] Posons que le premier en ait deux et l'autre une; ils jouent maintenant une partie, dont le sort est tel que, si le premier la gagne, il gagne tout l'argent qui et au jeu, savoir 64 pistoles ; si l'autre la gagne, ils sont deux parties à deux parties, et par conséquent, s'ils veulent se séparer, il faut qu'ils retirent chacun leur mise, savoir chacun 32 pistoles.

Considérez donc, Monsieur, que si le premier gagne, i lui appartient 64: s'il perd, il lui appartient 32. Donc, s'ils veulent ne point hasarder cette partie et se séparer sans la jouer, le premier doit dire: « Je suis sur d'avoir 32 pistoles, car la perte même me les donne ; mais pour les 32 autres, peut-être je les aurai, peut-être vous les aurez, le hasard est égal ; partageons donc ces 32 pistoles par la moitié et me donnez, outre cela, mes 32 qui me sont sures ». Il aura donc 48 pistoles et l'autre 16.

Posons maintenant que le premier ait deux parties et l'autre point, et ils commencent à jouer une partie. Le sort de cette partie est tel que, si le premier gagne, il tire tout l'argent, 64 pistoles ; si l'autre gagne, les voilà revenus au cas précédent auquel le premier aura deux parties et l'autre une.

Or, nous avons déjà montré qu'en ce cas il appartient à celui qui a les deux parties, 48 pistoles: donc s'ils veulent ne point jouer cette partie, il doit dire ainsi: « Si je la gagne, je gagnerai tout, qui et 64 ; si je perds, il m'appartiendra légitimement 48: donc donnez-moi les 48 qui me sont certaines au cas même que je perde, et partageons les 16 autres par la moitié, puisqu'il y a autant de hasard que vous les gagniez comme moi ». Ainsi il aura 48 et 8, qui sont 56 pistoles.

Posons enfin que le premier n'ait qu'une partie el l'autre point. Vous voyez, Monsieur, que, s'ils commencent une partie nouvelle, le sort en est tel que, si le premier la gagne, il aura deux parties à point, et partant, par le cas précédent, il lui appartient 56 ; s'il la perd, ils sont partie à partie donc il lui appartient 32 pistoles. Donc il doit dire : « Si vous voulez ne la pas jouer, donnez-moi 32 pistoles qui mes sont sures, et partageons le reste de 56 par la moitié. De 56 ôtez 32, reste 24 ; partagez donc 24 par la moitié, prenez en 12, et moi 12, qui, avec 32, font 44 ».

[11] We intend to try these activities out in the academic year 2010-11.

[12] This online course was conducted in Catalonia for the first time during last academic year, 2009-10, and supported by the Department Education of the Generalitat of Catalonia. The coordinator was Pere Grapí and the teachers were Agustí Camós, Pere de la Fuente, Pere Grapí, M. Rosa Massa, Fàtima Romero and Raimon Sucarrats.

Author index

✉ **Bächtold, Manuel**
ERES – LIRDEF – University Montpellier 2, France
manuel.bachtold@montpellier.iufm.fr

✉ **Bruneau, Olivier**
Université de Lorraine, Laboratoire d'Histoire des Sciences et de Philosophie – Archives Henri Poincaré, UMR 7117, Nancy, F-54000, France
olivier.bruneau@univ-lorraine.fr

✉ **Camós Cabecerán, Agustí**
Centre d'Història de la Ciència (CEHIC), Universitat Autònoma de Barcelona, Spain
acamos@pie.xtec.es

✉ **Cibulskaitė, Nijolė**
Lithuanian University of educational Sciences, Lithuania,
nci@takas.lt

✉ **Ferrière, Hervé**
Centre de Recherches et de Ressources pour l'Enseignement et la Formation, Institut Universitaire de Formation des Maîtres de Guadeloupe, Université des Antilles-Guyane, Centre F. Viète, Université de Bretagne Occidentale, France,
hferriere@iufm.univ-ag.fr

✉ **Garlatti, Serge**
Computer Science Department, TELECOM-Bretagne, France,
serge.garlatti@telecom-bretagne.eu

✉ **Le Gars, Stéphane**
Centre François Viète, Université de Nantes, Nantes, France,
stephlegars@free.fr

✉ **Gilliot, Jean-Marie**
Computer Science Department, TELECOM-Bretagne, France,
jm.gilliot@telecom-bretagne.eu

✉ **Grapí, Père**
Centre d'Estudis d'Història de la Ciència (CEHIC), Universitat Autònoma de Barcelona, Spain,
pere.grapi@uab.cat

✉ **Guedj, Muriel**
ERES – LIRDEF – University Montpellier 2, France
muriel.guedj@montpellier.iufm.fr

✉ **Guevara Casanova, Iolanda**
Ins Badalona VII, Badalona 08912 & Institut de Ciències de l'Educació de la Universitat Politècnica de Catalunya, Barcelona, Spain
iolanda.guevara@gmail.com

✉ **Heering, Peter**
Institute for Physics and Chemistry and its Didactics, Universitaet Flensburg, D-24943 Flensburg, Germany,
peter.heering@uni-flensburg.de

✉ **Kanellos, Ioannis**
Department of Computer Science, TELECOM-Bretagne, France,
ioannis.kanellos@telecom-bretagne.eu

✉ **Laubé, Sylvain**
Centre François Viète (EA 1161), Université de Brest, France,
sylvain.laube@univ-brest.fr

✉ **Lawrence, Snezana**
Bath Spa University, England,
s.lawrence2@bathspa.ac.uk, snezana@mathsisgoodforyou.com

✉ **Massa-Esteve, Mª Rosa**
Departament de Matemàtica Aplicada I., Universitat Politècnica de Catalunya, Spain,
m.rosa.massa@upc.edu

✉ **Pham-Nguyen, Cuong**
Faculty of Information Technology, Ho Chi Min Ville, Vietnam,
cuong.nguyen@telecom-bretagne.eu

✉ **Puig-Pla, Carles**
Centre de Recerca per a la Història de la Tècnica, ETSEIB, Universitat Politècnica de Catalunya, Spain,
carles.puig@upc.edu

✉ **Rebai, Issam**
Computer Science Department, TELECOM-Bretagne, France,
issam.rebai@telecom-bretagne.eu

✉ **Romero Vallhonesta, Fàtima**
Depart. de Matemàtica Aplicada I, Universitat Politècnica de Catalunya, Barcelona, Spain
fatima.romerovallhonesta@gmail.com

✉ **Sucarrats Riera, Raimon**
Centre d'Història de la Ciència (CEHIC), Universitat Autònoma de Barcelona, Spain
rsucarra@xtec.net

✉ **de Vittori, Thomas**
Laboratoire de Mathématiques de Lens, Université d'Artois, France
thomas.devittori@euler.univ-artois.fr